Transnational Environmental Governance

To Simen and Jildou

Transnational Environmental Governance

The Emergence and Effects of the Certification of Forests and Fisheries

---

Lars H. Gulbrandsen

Senior Research Fellow, the Fridtjof Nansen Institute, Norway

Edward Elgar

Cheltenham, UK • Northampton, MA, USA

Published by
Edward Elgar Publishing Limited
The Lypiatts
15 Lansdown Road
Cheltenham
Glos GL50 2JA
UK

Edward Elgar Publishing, Inc.
William Pratt House
9 Dewey Court
Northampton
Massachusetts 01060
USA

Paperback edition 2012

A catalogue record for this book is available from the British Library

Library of Congress Control Number: 2009941162

ISBN 978 1 84844 528 4 (cased)
ISBN 978 1 78100 710 5 (paperback)

Printed and bound in Great Britain by Marston Book Services Limited, Didcot

Contents

Figures, tables and boxes

FIGURES

TABLES

BOXES

Abbreviations

AF&PA	American Forest and Paper Association
ASI	Accreditation Services International
ATFS	American Tree Farm System
CARs	corrective action requirements
CBD	Convention on Biological Diversity
CDM	Clean Development Mechanism
CEPI	Confederation of European Paper Industries
CPET	Central Point of Expertise on Timber
CPPA	Canadian Pulp and Paper Association
CSA	Canadian Standards Association
CSD	UN Commission on Sustainable Development
DEFRA	UK Department for Environment, Food and Rural Affairs
ECOSOC	UN Economic and Social Council
EEZs	Exclusive Economic Zones
EMAS	EU Eco-management and Audit Scheme
EMS	environmental management system
FAO	UN Food and Agriculture Organization
FERN	Forests and the European Union Resource Network
FLEG	Forest Law Enforcement and Governance
FLEGT	Forest Law Enforcement, Governance and Trade
FLO	Fairtrade Labelling Organizations International
FSC	Forest Stewardship Council
G8	Group of Eight (Canada, France, Germany, Italy, Japan, Russia, UK, USA)
GASSDD	Guidance for the Assessment of Small-scale and Data-deficient Fisheries
GFTN	Global Forest and Trade Network
GHG	greenhouse gas
GRI	Global Reporting Initiative
HCVF	high conservation value forests
IAF	International Accreditation Forum
IATTC	Inter-American Tropical Tuna Commission
IFF	Intergovernmental Forum on Forests
IFIR	International Forest Industry Roundtable

IFOAM	International Federation of Organic Agriculture Movements
IPF	Intergovernmental Panel on Forests
ISEAL	International Social and Environmental Accreditation and Labeling
ISO	International Organization for Standardization
ITTA	International Tropical Timber Agreement
ITTO	International Tropical Timber Organization
IUCN	International Union for Conservation of Nature
IUU	illegal, unreported and unregulated fishing
MSC	Marine Stewardship Council
MTCS	Malaysian Timber Certification Scheme
NFPs	national forest programs
NGOs	non-governmental organizations
OECD	Organisation for Economic Co-operation and Development
PEFC	Programme for the Endorsement of Forest Certification (formerly known as Pan-European Forest Certification Scheme)
QACC	Questionnaire for Assessing the Comprehensiveness of Certification Schemes/Systems
REDD	reducing emissions from deforestation and forest degradation
SEPA	Swedish Environmental Protection Agency
SFI	Sustainable Forestry Initiative
SGS	Société Générale de Surveillance
SLIMF	small and low intensity managed forests
SSNC	Swedish Society for Nature Conservation
TBT	Technical Barriers to Trade
UNCED	United Nations Conference on Environment and Development
UNCTAD	United Nations Conference on Trade and Development
UNECE	United Nations Economic Commission for Europe
UNFCCC	United Nations Framework Convention on Climate Change
UNFF	United Nations Forum on Forests
WBCSD	World Business Council on Sustainable Development
WTO	World Trade Organization
WWF	World Wide Fund for Nature

Acknowledgements

This book grew out of my interest in the politics of environmental protection in forestry – an issue that has been part of my academic life for roughly a decade. My early work on forest politics investigated the influence of international environmental agreements and forest policy recommendations on Norwegian forest policies and forestry. Through this research, I discovered that non-state forest certification schemes seem to have had a greater impact on forest management and forestry practises than have the total of all the forest policy recommendations produced by intergovernmental bodies. Puzzled by this observation, I was eager to study the emergence and effectiveness of forest certification schemes in greater depth. How do forest certification schemes work? How do these schemes influence behavior? How did certification schemes emerge in the first place? Because it was modeled on forest certification, fisheries certification seemed to be a relevant case for comparison. As I undertook the research that became this book, I realized that transformation of the certification model as it was adopted in the fisheries sector could be a tremendously fruitful area for research. Thus the chapters on fisheries certification focus on knowledge acquired from certification experience gleaned within the forest sector – an understanding that paved the way for the world's leading wild-capture fisheries certification program. It was through this process that fisheries certification mimicked some of the features of forest certification, while strategically avoiding other features.

This book could not have been written without the help of many people. I am grateful to Arild Underdal and Olav Schram Stokke, who have supported and influenced this book project from its inception. Their encouragement and constructive comments have inspired me and helped me to focus on the key issues. Warm thanks go to Kristin Rosendal, who has stimulated my interest in social science research on environmental protection in forestry. I have also benefited greatly from discussions with and comments from many other colleagues throughout the writing process. Particular thanks go to Graeme Auld, Frank Biermann, Magnus Boström, Ben Cashore, Peter Dauvergne, Katarina Eckerberg, Alf Håkon Hoel, David Humphreys and Connie McDermott.

I am thankful to my colleagues at the Fridtjof Nansen Institute (FNI) for providing me with a stimulating and supportive working environment.

I could not imagine a more sustainable habitat than FNI in which to write a book like this one. But in spite of the nurturing intellectual environment that FNI provides, every researcher can benefit from leaving a safe sanctuary from time to time, in search of new habitats. I benefited tremendously from a six-month research stay as a visiting fellow at the John F. Kennedy School of Government, Harvard University, in the spring semester of 2007. My gratitude goes to Bill Clark and Nancy Dickson, co-directors of the Sustainability Science Program at Harvard's Center for International Development, for hosting me, commenting on my work and providing me with an excellent working environment. The stay was made possible through a scholarship from the Leiv Eiriksson program of the Research Council of Norway.

Thanks are also due to a number of commentators at various conferences and workshops and to reviewers of my work. Parts of the book draw on four of my articles: 'Overlapping public and private governance: can forest certification fill the gaps in the global forest regime?' in *Global Environmental Politics*, **4** (2), 2004, 75–99; 'The effectiveness of non-state governance schemes: a comparative study of forest certification in Norway and Sweden' in *International Environmental Agreements*, **5** (2), 2005, 125–49; 'Creating markets for eco-labelling: are consumers insignificant?' in *International Journal of Consumer Studies*, **30** (5), 2006, 477–89; and 'The emergence and effectiveness of the Marine Stewardship Council', in *Marine Policy*, **33** (4), 2009, 654–60.

I am grateful to Nina Colwill for her careful and professional language editing of the entire manuscript. I extend my thanks to Maryanne Rygg at FNI for her assistance with the editing of the original manuscript. I am also grateful to Felicity Plester and Bob Pickens at Edward Elgar for their support and perseverance and to Virginia Williams for her outstanding copy-editing of the manuscript.

I am indebted to my parents, Elisabeth and Rolf, for their never-ending support and intellectual encouragement. My brother Martin pursued an academic career in mathematics, and although I don't understand his mathematical proofs, I continue to enjoy our conversations about academic and nonacademic life. And, finally, I am tremendously grateful to Jildou for her love, care and inspiration, given in so many ways. Our son, Simen, made his appearance toward the end of this project. He made finalizing the book manuscript more challenging and more rewarding, and he serves as a constant reminder of our obligation to future generations to find effective ways of resolving global environmental problems.

Lars H. Gulbrandsen
Oslo
December, 2009

1. Introduction

A multitude of public and private organizations engage in international standard-setting processes. Some organizations establish technical standards to coordinate business or government behavior for a number of issue areas like the distribution of radio frequencies, international aviation, maritime classification and transportation, global communication systems, financial reporting and accounting, the size and shape of nuts and bolts and the like. Other organizations set standards for international games and sports, governing everything from international football events to the organization of local chess clubs around the world. Some organizations develop standards for voluntary information disclosure: two examples are the Global Reporting Initiative, a leading global standard in the field of nonfinancial reporting; and the Extractive Industries Transparency Initiative, a public-private initiative created to increase the transparency of payments made by companies in the extractive industries to host governments. Other organizations, like the chemical industry's Responsible Care, develop industry-wide codes of conduct to promote specific principles, norms and guidelines for environmentally responsible conduct. It is evident, then, that standard setting is neither a new activity nor a phenomenon limited to specific industries or sectors (Brunsson and Jacobsson 2000).

Over the past two decades, however, non-state actors have created a new type of institution for transnational environmental governance – in the shape of market-based certification programs. Various non-state certification schemes have emerged in response to perceived public policy failures, and have become particularly vibrant sources of rulemaking and governance. Sometimes referred to as the 'privatization of governance', non-state certification schemes have been launched in many sectors, ranging from forest and fisheries to eco-tourism and fair-trade initiatives (for example Cashore 2002; Honey 2002; Phillips et al. 2003; Conroy 2007; Raynolds et al. 2007). These programs typically establish environmental performance standards, as well as labor standards and other standards for socially responsible production. They go beyond voluntary codes of conduct and self-regulatory modes of governing, by involving the development of prescriptive standards, which require behavioral changes and independent verification of compliance (Cashore 2002). They also constitute governing arenas in which a wide range of stakeholders interact and agree upon rules and governance mechanisms (Bernstein and Cashore 2007). Producers participate in these certification schemes on a voluntary basis; as they are created and governed by non-state actors, there is no use of legal coercion to make producers sign onto the schemes. Rather, activists and advocacy coalitions use a range of strategies to convince, pressure or force producers to participate.

This book focuses on certification schemes in the forest and fisheries sectors, as initiatives in these sectors are among the most advanced cases of non-state rulemaking and governance. The Forest Stewardship Council (FSC) has served as a model for other certification schemes and arguably represents the most successful and well-known certification program to date. Although certification schemes were first developed through forestry initiatives, fisheries shared similar concerns: resource depletion, environmental degradation and insufficient governmental action. The Marine Stewardship Council (MSC) was formed by the World Wide Fund for Nature (WWF) and the multinational corporation, Unilever, which sought to model a fisheries sustainability program after the FSC. The certification model has spread to many other sectors as well, including sustainable tourism, park management, palm oil production, soy production and the marine aquarium trade. An examination of forest and fisheries certification can help us to understand the emergence and evolution of market-based governance programs and their effectiveness in addressing the problems or public policy failures that motivated their establishment.

This chapter proceeds as follows. The first section outlines the key research themes and research questions in the book. The second section provides a brief introduction to the literature on transnational environmental governance and non-state governance programs. The third section examines the features that render these programs different from other types of governance initiatives in the transnational realm. The chapter closes with an overview of the plan for the book.

RESEARCH THEMES AND RESEARCH QUESTIONS

The overall purpose of this book is to contribute to the understanding of an under-explored area of contemporary environmental politics: the emergence and effectiveness of non-state governance institutions in the shape of voluntary certification programs. The overarching research questions in this book are fundamental to political science. How can we explain institutional formation? How do institutions influence behavior?

Although we know a great deal about the formation and effectiveness of international regimes established by states, we still have limited knowledge about the emergence and effectiveness of transnational governance schemes developed by non-state actors such as non-governmental organizations (NGOs) and industry associations. This book focuses on three themes – the emergence, organization and effectiveness of non-state certification schemes – by examining three broad research questions:

- How can we explain the emergence of non-state certification programs in the forest and fisheries sectors?
- How do certain program designs emerge, and how and to what extent does program design influence standard-setting outcomes?
- What is the effectiveness of certification programs in resolving or ameliorating the problems that motivated their establishment?

Emergence

The origins of non-state certification in the forest and fisheries sectors serve as the first theme in this book. Forest and fisheries certification programs have become innovative venues for non-state rulemaking and governance. But why did certification programs emerge in these sectors; how did they evolve; and how have environmental NGOs, producers, consumers and states influenced non-state certification initiatives? More broadly, an examination of forest and fisheries certification can help us to understand the question of why non-state governance institutions increasingly supplement or supplant state-based, territorial government. This book explores the processes and mechanisms of the emergence and proliferation of non-state governance schemes. Although scholars have begun to identify the conditions that helped to establish certification programs in either the forest sector (Cashore et al. 2004; Bartley 2007; Pattberg 2007) or the fisheries sector (Auld 2007; Gulbrandsen 2009), there are few comparative studies of the emergence of certification in these two sectors. A comparative study of forest and fisheries certification and an analytical focus on processes and mechanisms can enrich the literature on non-state governance and private institutions. It can also enrich the literature on international environmental regimes, which lacks an empirical grounding in specific cases of non-state governance.

Organization

The second theme in this book is the organizational design of non-state certification programs. As discussed in this chapter, existing programs possess both commonalities and significant differences. Specifically, there are a number of ways to organize rulemaking processes, standard-setting activities and certification procedures. Three basic questions are addressed. How did certain program designs emerge? Did one particular design eventually emerge as the dominant and appropriate certification model? How and to what extent did program design influence standard-setting outcomes?

Non-state certification programs constitute governing *arenas* that assemble various stakeholders, regulate their interactions and provide opportunities for learning and for the mutual adaptation of behavior (Bernstein and Cashore 2007). In referring to an institution as a governing arena, we are interested in the access of actors to decision-making processes, their decision-making rights and their influence on decision-making outcomes. Governing arenas must have such mechanisms as decision rules and procedures for aggregating preferences into collective decisions. Standard setters must also decide which type of actors that should be allowed to participate in rulemaking, and what role they should play in the governance process.

In analyzing certification programs as governance arenas, this book examines how *constitutive* rules regulating the access, participation and decision-making rights of stakeholders influence the unfolding of the standard-setting process and the outcome of that process. Constitutive rules can be expected to influence standard-setting processes and the *regulative* rules being produced – the standards that regulate the behavior of certified producers (cf. Pattberg 2007, pp. 53–4). Industry domination in standard-setting bodies can be expected to result in industry-friendly standards, for example, whereas NGO domination can be expected to result in stricter social and environmental standards. In focus here is also the way in which non-state actors organize rulemaking and governance to create legitimacy for their actions and to enhance accountability. The creation of non-state certification programs can be seen as an effort by civil society organizations or industry associations to institutionalize accountability mechanisms beyond the nation state (Gulbrandsen 2008). Instead of simply replicating the traditional, territorial accountability structures in democratic states, these programs could create new tools and mechanisms that may prove more effective in holding producers to account than could traditional government regulations. The organizational focus of this book enables a detailed analysis of the emergence of certain program designs and their influence on standard-setting processes and outcomes.

Effectiveness

The third theme in this book is the effectiveness or performance of non-state certification schemes. The central question surrounding this issue is: How do we determine the effectiveness of certification programs as institutions of social and environmental governance? In order to answer this question, we need to clarify exactly what is meant by 'institutional effectiveness'. Following the most common definition of regime effectiveness, institutions of environmental governance can be considered effective if they contribute to the alleviation or resolution of the specific problems they address (for example Underdal 1992, 2002; Young and Levy 1999). Fisheries certification, for instance, was introduced to address grave problems in the fisheries sector – including overfishing caused by the overcapacity of fishing fleets and habitat destruction caused by wasteful fishing practices and vessel pollution. Forest certification was introduced to counter environmental degradation in forestry, caused by such practices as logging of old-growth forests, use of harmful pesticides and herbicides, clear-cutting of large areas, and drainage and ditching of forest wetlands. In this context, the effectiveness of a certification scheme could be operationalized as the degree to which the certification scheme modifies fisheries and forestry practices in ways that can resolve or alleviate these problems.

A distinction should be made between the direct effects of an institution and the broader consequences flowing from institutional-formation efforts (Underdal 2002, p. 5). As for the direct institutional effects, determining if the problem at hand can be solved under present certification schemes requires us to identify the causal mechanisms that mediate between certification schemes and changes in problem-relevant behavior, as well as the variables that influence problem-solving effectiveness. Specifying theoretically based and empirically grounded causal mechanisms is important for understanding the relationship between institutions and changes in problem-relevant behavior (Elster 1989; Young and Levy 1999). Detailed process tracing and case-study analysis of certification schemes can uncover when and under what conditions these mechanisms influence the behavior of certified producers.

In assessing broader consequences, I look beyond the certification instrument itself to discuss (1) unintended consequences of forest and fisheries certification, (2) sites of diffusion for the certification model, and (3) institutional interaction between certification programs and governmental and intergovernmental regulations. The first area for investigation is consequences not necessarily intended or anticipated by those who created forest and fisheries programs – consequences such as favoring large-scale over small-scale operations because of the benefits of economies of scale. The second area is the spread of the certification model across sectors and industries, from forests and fisheries to eco-tourism and fair-trade initiatives. Third is the interplay between public and private institutions governing natural resource use and protection. Because certification initiatives exist alongside existing international institutions and national laws and regulations, it is necessary to discuss interaction effects between public and private rulemaking and governance. Private and public institutions can reinforce each other's rules and enforcement capacities (positive interplay), but they may also disrupt or impede each other's effectiveness (negative interplay). In this context, it is interesting to examine certification initiatives to see if they tend to supplement or supplant traditional public policy regulations.

TRANSNATIONAL ENVIRONMENTAL GOVERNANCE

Traditional state-driven top-down governance approaches are being increasingly complemented by shared public and private authority, cooperative partnerships, voluntary standards, codes of conduct, and business self-regulation, prompting widespread claims that we are witnessing a shift from *government* to *governance* (Rosenau 1995, 2000; Rhodes 1996, 1997; Mol et al. 2000; Pierre 2000). In recent years, a literature on global governance and multi-level governance has emerged in opposition to the state-centric ontology of traditional international relations theorizing. Global governance approaches are used to capture and understand the myriad networks and steering arrangements in world affairs in the absence of an overarching authority at the international level (for example Rosenau 1997, 2003). Being depicted as an alternative to state-centered intergovernmentalist approaches, the concept of multi-level governance has been applied primarily to studies of European Union policy making and politics (for example Hooghe and Marks 2001). It also has relevance, however, for the study of non-state governance, in that it directs attention to the multiplicity of actors and networks engaged in policy making and enactment at different levels of authority and in various sectors (Bache and Flinders 2004).

Multi-level and global-governance scholars have in common their assertion that, as a result of globalization, centralization and supranational integration on the one hand, and localization, regionalization and fragmentation on the other, states must increasingly share rulemaking authority with subnational, transnational and supranational actors. They point to the alleged failure of international relations theory to capture adequately the effects of globalization and the increasing salience and impact of such non-state actors as environmental NGOs, social movement organizations and multinational corporations. Studies of global governance typically focus on nonhierarchical, network-based modes of governing, in which a range of actors are involved. Consequently, we should look beyond intergovernmental regimes to identify the central governance arenas and the key actors in a transnationalizing world.

A growing literature has explored the influence of transnational advocacy coalitions around the world. The most profound change between the earlier period of corporate criticism and the present climate is arguably the explosion of transnational activist and NGO networks (Haufler 2001). Whereas activism in the past was limited to the domestic arena, activists today organize across national boundaries, convening stakeholders in a number of countries to pressure companies or governments (for example Risse-Kappen 1995; Wapner 1996; Keck and Sikkink 1998; McAdam et al. 2001). Activists and NGOs have traditionally targeted governments to force them to make companies accountable by enacting laws and regulations. In the era of globalization, transnational activist networks are using increasingly sophisticated methods to hold companies accountable. They seek to mobilize consumer sentiment through calls for boycotts, and they seek to mobilize investors through socially responsible investment funds and appeals to shareholders (Haufler 2001). They make effective use of the Internet, and orchestrate coordinated media campaigns across the globe to force companies to reform policies (cf. Spar 1998).

According to the notion of 'political consumerism' (Micheletti 2003; Micheletti et al. 2004), ethically and politically motivated consumers may force policy reform or persuade producers to abandon questionable practices through their choice of producers and products. Besides making political and ethical purchases, citizens can boycott particular brands and products from companies with economic, social or environmental practices that fall from public favor. Social movement organizations facilitate political consumerism by praising or damning industry practices, mobilizing consumers and advising companies on tapping into potential consumer demand. In short, the global marketplace is not only a place for economic transactions but also a site of political agitation by social movement organizations, interest groups and consumers.

The rise of transnational regulation and non-state governance institutions in world politics is a topic of great interest to an increasingly large number of scholars across many disciplines. Scholars have begun to explain the conditions that helped to establish non-state institutions in world politics. In a much-cited volume on the increasing salience of 'private authorities' in international affairs, Cutler et al. (1999, p. 16) note that the literature on international regimes has generally underestimated the role of non-state actors, maintaining that these actors are 'increasingly engaged in authoritative decision-making that was previously the prerogative of sovereign states'. Their edited volume includes case studies of standards for such issues as online commerce, telecommunication and information technologies, maritime transport, intellectual property rights, business self-regulation and the governance of international mineral markets. Cutler et al. convincingly argue that transnational private-sector governance schemes increasingly supplement governance arrangements involving states, but they do not assess the effectiveness of such schemes. Moreover, by focusing only on cases in which business creates the rules, they exclude cases in which non business interests hold or compete for rulemaking authority (Cashore 2002).

The expansion of private authority in international affairs is further explored in a volume edited by Hall and Biersteker (2002), which includes cases on the role of private sector markets (market authority), civil society and transnational regulation (moral authority) and transnational organized crime (illicit authority). One chapter by Lipschutz and Fogel (2002) examines the emergence of transnational environmental regulation. They remain skeptical of the capacity of civil society regulations to provide quick fixes to problems of ineffective transboundary environmental governance and lack of democratic accountability in a transnationalizing world. Similarly, in a study of self-regulation in industry, Haufler (2001) concludes that such regulations are unlikely to fill the policy void if public regulation of social and environmental problems is weak or absent. She attributes the spread of private standards for environmental regulation, worker rights and data privacy to public pressures from activist campaigns and the threat of public policy regulations. Yet, like Lipschutz and Fogel (2002), Haufler does not assess variation in the effectiveness of business and civil society regulations (cf. Vogel 2008, p. 263).

In recent years, political science and sociology scholars have examined the emergence of forest certification programs (Bartley 2003, 2007; Cashore et al. 2004; Elliott 1999; Gulbrandsen 2004; Klooster 2005; McNichol 2002; Overdevest 2004, 2005; Pattberg 2007). Cashore et al. (2004) have conducted the most comprehensive and academically rigorous social science study on the emergence of forest certification. They explained the variation in support for FSC among forest owners and forest companies in British Columbia, Canada; the USA; the UK; Germany; and Sweden. Their analysis offers a number of valuable insights upon which the analytical framework in this book draws, but they did not investigate the impact of forest certification. Nor did they examine certification initiatives in sectors other than forestry.

This book contributes to the literature on certification programs and the emergence of private authority. It also expands on the evolving political and sociological scholarship regarding green consumerism (Boström and Klintman 2008; Micheletti 2003; Micheletti et al. 2004; Oosterveer 2005) and the legitimacy and accountability of non-state certification programs (Bernstein and Cashore 2007; Boström 2006a, 2006b; Cashore 2002; Dingwerth 2007; Gulbrandsen 2008). And finally, the book contributes to the emerging literature on the problem-solving effectiveness of forest and fisheries certification programs (Auld et al. 2008; Cashore et al. 2007; Gulbrandsen 2009; Ward 2008).

Non-state governance institutions are generally seen as emerging in response to globalization processes and transboundary problems that states have been unable or unwilling to resolve themselves (for example Djelic and Sahlin-Andersson 2006; Pattberg 2007). Indeed, many of these institutions have been created with little or no involvement of states or traditional international organizations like the UN or the World Bank. In the apparel products field, for example, NGOs created labor standard certification schemes to address sweatshop labor practices, child labor and other human rights violations (Bartley 2005). For other issues, such as trading in coffee, bananas and a wide range of other commodities, NGOs have taken the initiative to create fair-trade labelling schemes to guarantee marginalized producers in developing countries a fair minimum price for their products and to improve their working conditions (Raynolds et al. 2007). Realizing that traditional boycott campaigns often failed to encourage companies to opt for more sustainable production practices, the creation of non-state certification schemes was, in many cases, an NGO effort to find new ways to influence corporate conduct.

This shift from transnational activism and boycotts to transnational rules and governance programs is precisely the focus of this book. Although the activities of companies have resulted in a wide range of social and environmental problems, such as child labor, ozone depletion, climate change, deforestation and overfishing, companies are increasingly seen as part of the solution to these problems. In the new climate of voluntary policy making, NGOs assume new roles through strategic engagement with companies and industry associations. It means that a broad range of actors engage in rulemaking processes – often in governance arenas located outside traditional channels of political influence. The next section examines the characteristics of the governance tool in focus in this book: non-state certification schemes.

APPRAISING NON-STATE CERTIFICATION SCHEMES

The purpose of delineating the features of non-state certification schemes is to understand better their uniqueness as a new form of social and environmental governance. According to Cutler et al. (1999, p. 19), three features of 'private authority' render their rulemaking authority distinct: first, those subject to the rules being made by private actors must accept the rules as legitimate. Second, there must be a high degree of compliance with rules and decisions being made by private actors. Third, private-sector actors must be empowered either explicitly or implicitly by governments and international organizations granting them the right to make decisions for others.

This latter assumption has been questioned by Cashore (2002), who argues that it is precisely the lack of government delegation of rulemaking authority that is one of the defining features of market-based certification programs, or what he calls 'non-state market-driven' governance. Although states may influence non-state governance systems, they do not use their sovereign authority to require compliance with rules. Governments can act as traditional interest groups attempting to influence rulemaking in non-state governance systems; they can act like any large market player by initiating procurement policies; and they can sometimes act like producers seeking certification of government-controlled operations. But non-state market-driven governance systems do not derive rulemaking authority from states. Of course, governments can and sometimes do use their sovereign authority to require adherence to standards developed by non-state actors. In this event, however, the logic of market-driven support no longer explains why producers adopt and comply with the standards, and it is, therefore, no longer a case of non-state market-driven governance (Cashore 2002, p. 510).

According to Cashore, the logic of market-driven support means that authority granted to non-state market-driven governance schemes emanates from the market's supply chain. Producers and consumers along the supply chain make their own evaluations about whether or not to grant authority to these schemes. The market's supply chain provides the incentives through which evaluations of support occurs. Compliance incentives in the form of a promise of price premiums, market access or prevention of boycott campaigns are created up and down the commodity supply chain. In this way, non-state market-driven governance systems aim to ameliorate social and environmental problems through the *reconfiguration* of markets (Bernstein and Cashore 2007, p. 350). Unlike business coordination standards, these governance systems seek to create incentives for producers to address problems that they would otherwise have little incentive to address. This characteristic of non-state market-driven governance systems distinguishes them from most other types of private and public-private governance arrangements (Cashore 2002, pp. 511–13).

Other scholars have questioned this conceptualization of certification as a non-state market-driven mode of governance, highlighting the problematic nature of the public-private distinction (Boström 2003; Meidinger 2006; Tollefson et al. 2008) Yet, the dominant tendency in the literature has been to characterize certification programs as non-state or private regulation (Bartley 2003, 2005; Cashore et al. 2004; Klooster 2005; Lipschutz and Fogel 2002; Pattberg 2007; Vogel 2005, 2008). While acknowledging the difficulties with characterizing certification programs as 'non-state' and 'market-driven', I find Cashore's definition helpful as a point of departure for distinguishing these programs from other governance experiments.

With the characteristics of non-state certification programs briefly reviewed, it is now useful to describe their specific attributes in greater detail. Social and environmental certification programs generally comprise the following features (cf. Auld 2007, p. 4; Meidinger 2006; Bernstein and Cashore 2007; Cashore 2002):

1. They have standards for regulating the social and/or environmental impact of production processes. Yet, the stringency and scope of these standards vary across programs. Some programs have relatively stringent standards for regulating both the social and environmental impact of production within a sector or industry. Other programs have flexible and discretionary standards for regulating only specific aspects of the production process. The size and type of target groups (producers) for the standards also varies. Some programs seek to approve only the social and environmental frontrunners within an industry, whereas other programs seek industry-wide adoption of standards.
2. They have mechanisms to verify compliance with the standards and create consequences for noncompliance. Verification of compliance usually involves a certification procedure in which auditors assess whether or not producers meet the standards. Producers that pass the inspection of on-the-ground practices are awarded a certificate attesting to compliance. If producers fail to correct serious breaches of the standards, they risk the loss of their certification. The scope of the auditing process and the consequences for failing to comply vary among programs, however. Although most programs involve on-the-ground inspections, the number and types of issues addressed by auditors vary by program. Depending on the seriousness of instances of noncompliance, but also on the rules of the certification program, the consequences for failing to comply range from minor or major requests to correct practices on the one hand, to revocation of the certificate on the other.

3. They have rules for accreditation of third-party certification bodies (certifiers). Accredited certifiers usually conduct on-the-ground inspections and monitoring of producer practices, although certain programs involve only an internal verification process. Initially, accreditation of certifiers was often conducted within certification programs; but over time, it has, in many cases, become outsourced to independent accreditation organizations. These organizations accredit certifiers in accordance with the requirements established by the certification program.
4. They have governance bodies and rules for regulating membership in the program, decision making, standard-setting activities, complaints and dispute resolution. Membership rules and governance structures vary by program; some have membership and voting rules that are favorable to industry and business interests, whereas others have rules that balance decision-making powers among environmental, social and economic interests. There is also significant variation in the rules for standard-setting activities, handling of complaints and dispute resolution.
5. They have tracking requirements for following products originating from approved operations through to the end consumer, as well as logos or labels that can be used on the products. Tracking requirements, type of labels and rules for label use vary by program.

This list of attributes demonstrates that although there are many differences in program design, social and environmental certification programs have several features in common (Auld 2007). Based on these commonalities, we can identify cases of non-state governance that clearly fall outside this new form of governance. This delimitation is important, because scholars of transnational governance systems tend to conflate the new type of certification programs and a wide range of other governance experiments, ignoring the unique and innovative features of the certification programs (Cashore 2002). Certification initiatives in forestry (FSC), fisheries (MSC), the marine aquarium trade (Marine Aquarium Council), coffee production (Fairtrade Labelling Organizations International) and the apparel industry (Fair Labor Association) are examples of full-fledged social and environmental certification programs (Cashore et al. 2004). By contrast, a number of voluntary codes of conduct, sustainability reporting schemes and corporate social responsibility initiatives (see Vogel 2005) *do not qualify* as such programs. The Global Reporting Initiative (GRI), for instance – a leading global standard for non-financial business reporting – does not qualify because it does not involve mandatory verification of compliance with performance-based certification standards. How to use the GRI guidelines is a decision made within the firm, and reports are not certified by GRI or by accredited certification bodies (Dingwerth 2007, p. 107). Similarly, the UN Global Compact program – a set of ten universal principles in the areas of human rights, labor, the environment and anti-corruption – is not monitored, verified or enforced. Rather, it is dressed in the business-friendly language of dialogue, partnership and voluntarism, and it relies on public transparency rather than policing of the principles (Garsten 2008).

Generic environmental management system standards, such as ISO 14001 and the Eco-management and Audit Scheme of the European Union do not qualify either, because they are hybrid public-private schemes that do not involve verification of on-the-ground compliance with performance-based standards. Instead of prescribing environmental performance objectives, a management-system-based scheme focuses on organizational process design to meet internally established environmental objectives and continually improve performance. It typically requires organizations to establish and implement an environmental policy or plan, review progress through systematic auditing and correct problems (Coglianese and Nash 2002), but the performance level to which they will aim is decided within the organization (Krut and Gleckman 1998; Clapp 1998). These examples should suffice to illustrate that non-state environmental and social certification schemes represent a new form of governance, which differs significantly from the increasing number of voluntary codes of conduct, reporting programs and corporate social responsibility initiatives around the world. Table 1.1 shows some of the most influential non-state certification programs.

STRUCTURE OF THE BOOK

The next chapter sets out the theoretical and analytical framework of the book. Corresponding to the broad research questions of the study, I outline factors that are likely to influence the emergence of non-state certification schemes, the unfolding and outcome of their standard-setting processes and their effectiveness. I also discuss case selection and methodological considerations in conducting this study.

The two following chapters focus on the emergence, evolution and effectiveness of forest certification around the world. The focus in Chapter 3 is on the formation and evolution of forest certification schemes in Europe, North America and elsewhere. Particular attention is given to the competition between FSC and producer-dominated programs and the way this competition influenced the evolution of forest certification. Chapter

Table 1.1 Examples of non-state certification programs

	Origin	Initiators	Policy goal
Forest Stewardship Council	1993	Broad coalition of environmental NGOs and socially concerned companies	Environmentally and socially responsible forestry practices
Rainforest Alliance Certification	1993[1]	Rainforest Alliance (an NGO)	Sustainable farming through certification of a range of tropical commodities
Sustainable Forestry Initiative	1994[2]	American Forest and Paper Association	Sustainable forest management
Marine Stewardship Council	1997	World Wide Fund for Nature and Unilever	Environmentally responsible fishing practices
Social Accountability International	1997	Council on Economic Priorities (an NGO)	Protect workers' rights and improve working conditions
Fairtrade Labelling Organizations International	1997[3]	Broad coalition of NGOs and consumer groups	Guarantee developing country producers fair price, improve working conditions
Marine Aquarium Council	1998	Environmental NGOs, aquarium industry, public aquariums and hobbyist groups	Conserve marine ecosystems through promotion of responsible aquarium trade
Programme for the Endorsement of Forest Certification	1999	European forest owner associations	Sustainable forest management
Fair Labour Association	2001[4]	Industry, Clinton administration, consumer and labor rights organizations	End sweatshop conditions in factories
Roundtable on Sustainable Palm Oil Certification System	2007[5]	World Wide Fund for Nature and Unilever	Promote growth and use of sustainable palm oil

Table 1.1 (*continued*)

	Origin	Initiators	Policy goal
Aquaculture Stewardship Council	2010	World Wide Fund for Nature with participants of the Aquaculture Dialogues	Sustainable fish farming

Notes:
1. 1993 was the year the Rainforest Alliance certified the first two tropical farms under its agricultural certification program.
2. Initially an industry code of conduct with mandatory self-reporting for members of the American Forest and Paper Association, the Sustainable Forestry Initiative provided for voluntary third-party verification in 1998.
3. Fairtrade Labelling Organizations International united 15 separate initiatives.
4. 2001 was the year the Fair Labour Association established an independent auditing system.
5. A certification system for sustainable palm oil was launched at the fifth roundtable meeting in 2007.

Sources: Bernstein and Cashore (2007) and author's research.

4 reviews and synthesizes what we know about the effectiveness of forest certification. In addition to considering such measures as the forest area certified and the proportion of certified to uncertified forests, I examine patterns of adoption, market penetration and the impacts of on-the-ground auditing.

Comparing FSC and a landowner-dominated competitor, Chapter 5 examines the emergence and effectiveness of forest certification in Sweden and Norway. These countries stand out as particularly suited for comparison; they share many similarities, yet differ in the structure of their forest industries. In this chapter, I examine the ways in which this difference influenced forest certification choices and the subsequent evolution of certification programs.

Like the chapters on forest certification, the two chapters on fisheries certification focus on program origins, design and effectiveness. Chapter 6 examines the way the certification model was exported from the forest sector to the fisheries sector, detailing the origins and evolution of the MSC. Particular attention is paid to the development of the program, and how it was influenced by early choices of program features. Beginning with a review of patterns of adoption, Chapter 7 examines the effectiveness and environmental impact of fisheries certification. In this chapter, I also seek to explain patterns of adoption and discuss criticism of fisheries certification.

Chapter 8 looks beyond the certification instrument to discuss the institutionalization of multi-stakeholder governance programs and sites of diffusion for the certification model. The chapter details the role of environmental NGOs, certifiers and other policy entrepreneurs in spreading the certification model across sectors and industries – among them, the marine aquarium trade, sustainable farming and eco-tourism. This chapter closes with a discussion of the challenges for social and environmental certification programs.

In Chapter 9, I summarize the findings, discuss their relevance for the study of global environmental governance, reflect on some of the wider lessons from this study of forest and fisheries certification, and suggest directions for further research. My primary goal in this book is to provide the reader with a better understanding of the origins and evolution of non-state governance programs and the ability of these programs to address some of the most pressing global environmental problems facing humankind today.

2. Non-state governance: an analytical framework

The presentation in Chapter 1 demonstrates that forest and fisheries certification programs represent a new form of environmental and social governance. Analyzing the emergence and effectiveness of these programs requires the development of an analytical framework applicable across cases. Such a framework must facilitate a comparison of the conditions that influence the origins and effectiveness of forest and fisheries certification. The analytical framework must also be applicable to other cases of non-state governance in order to allow for broader comparisons and an expansion of the generalizability of the conclusions.

This chapter proceeds in three steps. The first section reviews theories that shed light on the causal mechanisms that help to explain institutional formation and effectiveness. The second section develops an analytical framework for examining the conditions under which non-state governance schemes are likely to emerge and influence the behavior of target groups. These conditions are identified in two ways: on the basis of extant work on private authorities and non-state governance institutions and inductively, from preliminary research on the cases. The third section discusses the methodology and material that guided and informed the analysis.

TWO THEORETICAL APPROACHES TO STUDYING INSTITUTIONS

Although the focus of this book is on non-state governance schemes, I share a research interest in exploring institutional formation and effectiveness with students of international regimes and organizations established by states. The analytical framework thus draws on two well-established theoretical perspectives in the social sciences: rational institutionalism and sociological institutionalism. Although these perspectives are sometimes said to be incompatible in terms of ontological and epistemological premises, I argue that insights from rational and sociological institutionalism can be combined to examine and understand institutional formation and consequences. To draw upon insights from these theoretical traditions resonates well with recent efforts to bridge the gap between the rationalist and constructivist literatures on international relations (for example Adler 1997; Checkel 1997, 2007; Finnemore and Sikkink 1998; Risse et al. 1999; Fearon and Wendt 2002), and to draw upon sociological institutionalism in the study of international organizations (Finnemore 1996; Barnett and Finnemore 1999, 2004). Indeed, accounts of international regime formation and effectiveness often draw on insights from both rational and sociological institutionalism (Stokke 1997; Young 1999; Young and Levy 1999), as could accounts of non-state governance institutions.

Rational Institutionalism

Given the focus on *transnational* institutions in this study, the rationalist strand of regime theory stands out as one particularly useful approach for examining institutional formation and institutional effects. To be sure, there are several theoretical approaches within regime theory that are not dissimilar to broader theoretical approaches within the field of international relations theory. For instance, Hasenclever et al. (1997) argue that one can differentiate among interest-based, power-based and knowledge-based theories of international regimes. For the purpose of clarity, however, I am drawing on insights from the mainstream interest-based or rationalist strand of regime theory in this section in order to explore non-state institutional formation and consequences. Whereas state-centric, power-based (realist) accounts of international regime formation and effectiveness seem less relevant for the study of non-state regimes, I return to some of the insights from knowledge-based (constructivist) regime theory in the next section on sociological institutionalism.

Partly because of the different approaches taken to study regimes, there is some disagreement over how to define and delineate regimes (see for example Young 1986; Levy et al. 1995). The most commonly cited definition is probably Krasner's (1982, p. 186) specification of regimes as 'sets of implicit or explicit principles, norms, rules and decision-making procedures around which actors' expectations converge in a given area of international relations'. International environmental regimes typically include a core treaty, such as a framework convention, supplemented by one or several protocols, although they may also be based on 'soft law' agreements (Levy et al. 1995, p. 274). Prototypical examples of international environmental regimes include the ozone layer protection regime, based upon the Convention for the Protection of the Ozone Layer (1985) and upon the Montreal Protocol on Substances that deplete the Ozone Layer (1987); and the climate change regime, based upon the Framework Convention on Climate Change (1992) and the Kyoto Protocol (1997).

The assumption that states are the key actors in world politics and the focus on institution building and effects separate the literature on international regimes from the literature on multi-level governance and global governance. Cutler et al. (1999, p. 14) note that although the regime literature has remained 'stubbornly state centric' in its conceptual and empirical focus, 'the definition itself, and its utility in explaining certain forms of cooperation, does not require the relevant actors to be states'. Indeed, there are several similarities between regimes established by states and private regimes or governance schemes established by non-state actors. Being issue-specific within clear spatial and functional boundaries and with authority to enforce or facilitate compliance in their specific concerns, both public and private regimes seek to influence the behavior of their members. In essence, mainstream, rationalist regime theory is a theory about *voluntary collaboration* among actors to create mutually beneficial institutional arrangements in order to achieve some common goals. The attractiveness of applying insights from rationalist regime theory to analyzing non-state governance lies, in particular, in its focus on issue specificity, institutional formation, causal consequences and problem-solving capacity (cf. Stokke 1997; Young and Levy 1999; Miles et al. 2002).

From the perspective of structural realism, regimes are seen as epiphenomena that mirror and never change the fundamental configurations of power and interests in world politics (Strange 1982; Mearsheimer 1995). According to Waltz (1979), we must distinguish between institutions and what he calls *ordering principles*; the ordering principle of an anarchical international system means that the only international institutions that can be built and sustained are those based on consensual cooperation or hegemonic coercion. In response to the relatively pessimistic implications of structural realism, regime theorists have set out to demonstrate that institutions have causal autonomy and that they are not merely a reflection of configurations of power and interests in world politics. One of the main claims of the interest-based perspective is that regimes may change the utility that actors assign to behavioral options within an issue area. Unlike structural realists, regime analysts claim that once established, regimes may have significant behavioral consequences for their members – independent of underlying power structures. For example, an effective regime may reduce the risk of unilateral defection by increasing the costs of breaching certain principles, norms and rules, and/or by increasing the benefits of complying. More generally, regime theorists argue that because regimes enhance reciprocity, reduce barriers to mutually beneficial collaboration and are resilient to changes in the configurations of interests and power structures among states, they cannot be dismissed as epiphenomena in international relations (for example Keohane 1984, 1993; Young and Osherenko 1993; Haas et al. 1993).

Regime theorists in the rationalist tradition treat state interests as exogenously given. Many analysts conceive of states as unitary rational actors, although some also look at the influence of domestic interest groups. In an international society characterized by 'complex interdependence' (Keohane and Nye 1977), states have mutual interests in a variety of issue areas such as security, energy policies, policing, trade, monetary policies, sustainable resource management and the environment. Efforts to provide a public good (a good that cannot be denied to anyone once it is provided) through collective action always involve the risk of free riding by actors who do not share the costs of obtaining the good, but reap the benefits (Olson 1965). According to Olson (1965), if there were no penalties for failing to contribute to the realization of the public good, it would not be in the self-interest of rational, utility-maximizing actors to contribute to its realization, even though all actors would benefit from it.

Although international relations are beset with collective action problems, the relatively small number of states in the world decreases problems with collective action and enhances the likely success of collaboration (Keohane 1984, p. 77). Keohane's functional or contractualist theory of international regimes explains regime formations as, *inter alia*, efforts to resolve collective action problems and provide mutual goods by enhancing reciprocity and certainty about future interactions and reducing transaction costs and other barriers to mutually beneficial collaboration (Keohane 1984, 1989, 1993). Using game theory, Axelrod (1984) has demonstrated that collaboration among utility-maximizing actors can emerge as a result of repeated interactions over time. And, opposing Olson's (1965) relatively pessimistic view on collective action, Ostrom (1990) has argued that common-pool resources, such as inshore fisheries and communal forests, can be managed by common property regimes if they are properly designed.

The question then becomes: *How* do institutions produce effects? Regime theorists have tried to answer this question by tracing processes that mediate between the institutions and particular outcomes (Stokke 1997; Young and Levy 1999). Such process tracing is often guided by the specification of one or more *causal mechanisms* that are believed to link institutions and behavioral change. Whether we study international regimes established by states or non-state institutions for environmental governance, our task would then be to specify the ways in which the institutions may contribute to problem-solving behavioral adaptations. According to interest-based regime theory, the principle function of international regimes is to *restructure incentives* by increasing the benefits of participation and compliance with rules and by adding costs to defection (Barrett 2003). Similarly, certification schemes may influence the cost-benefit calculations of utility-maximizing companies by creating opportunities to profit from market demand for products flowing from sustainable resource-management practices.

One may expect that non-state actors, like states, agree on *coordination* standards to resolve coordination problems, decrease uncertainty and reduce transaction costs. For standards organizations to form in the first place, actors must perceive that coordination will serve their interests and that the achievement of any benefit (whether individual or collective) is contingent upon mutual action. Producers, firms and other market actors could therefore be expected to contribute to institutional formation and participate in those institutions to increase utility. In business coordination situations, in which actors are indifferent about where to coordinate behavior, all actors profit from collaboration and nobody profits from defection. Once established, an industry code of conduct or standard may be adopted by all companies and could in one sense be regarded as a collective good for the industry. Examples of coordination standards are international aviation safety standards like those established by the International Civil Aviation Organization, rules pertaining to the use of sea lanes created by the International Maritime Organization or global communication standards like those established by the Internet Corporation for Assigned Names and Numbers. As long as actors are indifferent about where to coordinate and are able to communicate, agreeing on such standards is relatively easy (cf. Axelrod 1984; Keohane 1984; Snidal 1985; Young 1999).

Because environmental and social reputations may reflect on the industry as a whole – not merely on individual firms (Gunningham and Rees 1997) – industry associations often adopt industry codes of conduct in order to demonstrate the high level of responsibility they assume for their operations, to protect the reputation of their industry and to provide credible information to consumers (cf. Klein 1997; Spar 1998). The collective action problem is reduced by the fact that companies often participate in industry associations and are able to monitor each other's behavior. An industry response of this kind occurred when the US chemical industry developed the Responsible Care code-of-conduct following the 1984 Bhopal Disaster in India, in which the accidental release of 40 tonnes of toxic gas from a pesticide plant owned by the US company Union Carbide killed several thousand people.

There is, however, a fundamental difference between business coordination standards and *performance-based* certification standards. Once coordination standards have been established, actors would have no incentive for cheating (cf. Young 1999, p. 27). Because all companies benefit from adopting business coordination standards, they pose no puzzle for compliance. By contrast, performance-based standards require companies to undertake costly behavioral changes that they otherwise would not be required to implement (Cashore 2002). Why, then, do profit-maximizing companies adopt and comply with performance-based standards on a voluntary basis? The interest-based strand of regime theory, argues that states – as rational, unitary actors – may and often do act within the constraints of rules for reasons of material self-interest and utility (see for example Hasenclever et al. 1997). It can be assumed that companies may also choose to adopt and comply with voluntary standards based upon rational-calculative decisions because compliance is expected to reduce costs or generate net benefits in the long term. We may expect that effective non-state certification programs, similar to regimes established by states, restructure incentives by increasing the benefits of standard adoption and penalizing defection. As explained by Cashore (2002), incentives for participation in market-based certification programs are created up and down global commodity chains. Such incentives can take the form of the promise of price premiums on certified products, greater market access or prevention of boycott campaigns. In short, the principal function of certification programs is to create rules and governance arrangements that contribute to a realignment of incentives governing resource management and use.

As noted, penalizing noncompliance will not be important in pure coordination situations, but will be essential in cooperation situations. A purely utility-maximizing actor could benefit from adopting performance-based standards and *not* complying with them, given that noncompliance is not detected or does not have negative consequences (Young 1999; see also Snidal 1985). It must be possible to detect noncompliance, therefore, and compliance must be enforced (Cashore 2002). The principal tool for monitoring and enforcing compliance in certification schemes is regular third-party auditing of practices. Companies that comply with the certification standards are rewarded with a certificate that attests to sustainable management practices. Companies that do not comply with standards risk the penalty of having their certificate suspended. From the perspective of rational institutionalism, compliance with non-state standards must, in principle, be monitored and enforced.

Sociological Institutionalism

From the perspective of sociological institutionalism, institutionalized norms in the environments of organizations define appropriate and inappropriate behavior, prescribe and proscribe courses of action and legitimate particular organizational forms (Meyer and Rowan 1977; DiMaggio and Powell 1983, 1991; Scott 2001). Organizations adopt a certain language and certain procedures because the actions of an entity must be acceptable or appropriate within a certain institutional framework. In early neoinstitutional work, organizations are said to reflect – and never to transform – institutionalized norms and values in the environments and systems in which they are situated. According to sociologists Meyer and Rowan (1977), formal organizations derive their form and function from institutionalized social orders. The adoption of certain formal structures is seen to be the result of the traveling and spread of *rationalized myths*. Myths are widely held belief systems and cultural frames that are imposed upon or adopted by organizations. They are rationalized because they prescribe certain ways of organizing and proscribe other ways of organizing to accomplish a given end. Organizations adopt rationalized myths and must reflect institutionalized social orders in order to be granted legitimacy from salient constituencies in their environments.

The effect of rationalized myths on institutional formation is organizational imitation and convergence. Organizations derive their form, not from instrumental efficiency, but rather from institutionalized norms and values in the environments in which they are situated. Because organizations are reflections of rather than creators of underlying structures, there are no autonomous causal effects from the organizations. In this perspective, the formal organization may be crucial for legitimizing behavior, but does little to change the rules of the game and underlying social orders. Similarly, constructivist regime theorists stress that international regimes are embedded in and molded by broader normative contexts (for example Ruggie 1983; Kratochwil and Ruggie 1986). According to this view, regimes are not so much creators of international norms and practices as they are reflections of underlying normative structures and social orders (Stokke 1997). Strong versions of this argument hold that underlying normative structures are fully determinate of regime design and social practices, whereas more moderate versions hold that normative structures are important sources of legitimization, but not full determinants of formal structure. In the latter view, there is room for both agency and transformation of underlying social orders. According to Conca (2006, p. 69):

> If the normative order of international relations is powerful without being fully determinate – authoritative but not hegemonic – then specific struggles to craft the rules, norms, and institutions of global environmental governance could yield institutional forms other than the statist, territorialized, functional-rational institutional form. Studying these struggles may shed light on whether a richer array of institutional forms than we can imagine exist in practice.

Sociological institutionalists subscribing to the notion of powerful and fully determinate institutional environments have long struggled with explaining *institutional change*. If organizations merely reflect deeper layers of social orders or configurations of power and interest, how can we explain why and how institutions emerge, evolve and sometimes die? Part of the answer is given by organizational field-level analyses (for example DiMaggio and Powell 1983; DiMaggio 1991), which shows that even though institutional environments are important, organizations can themselves be agents of change and transform the fields in which they are situated. Institutional theorists have developed the concept of *organizational field* to isolate for analysis 'a collection of interdependent organizations operating with common rules, norms, and meaning systems' (Scott 2003, p. 130). According to DiMaggio and Powell (1983, p. 148), an organizational field comprises 'those organizations that, in the aggregate, constitute a recognized area of institutional life'. Highly institutionalized fields are characterized by sets of rules and practices that are taken for granted. The character of an organization's embeddedness in a field shapes the organizational arrangements, procedures and strategies that are perceived to be legitimate. As explained by Scott (2003, p. 130), in field-level analyses, 'organizations are treated as members of larger, overarching systems exhibiting, to varying degrees, structure and coherence'.

Organizations in a specific field may not be linked by direct interactions, but they operate in the same realm and under similar conditions and therefore exhibit similar structural characteristics (Scott 2003, p. 130). Unlike the notion of relatively fixed institutional environments, a field-level perspective allows us to observe not only the influence of common norms, rules and meaning systems, but also the disappearance of some organizational types and the emergence of new forms. Whereas early neoinstitutional work tended to see organizations as adapting rather passively to rationalized myths, more recent work has demonstrated that organizations adapt and transform myths and innovate to create institutional change (for example Sahlin-Andersson 1996; Hoffman 1999; Brunsson and Jacobsson 2000; Sahlin-Andersson and Engwall 2002).

A sociological account of regime formation would contend that specific *organizational carriers* are agents of institutional change in organizational fields. International organizations, NGOs, business consultants and activists are said to constitute networks with a certain culture and significant influence on the formation, transformation and flow of organizational ideas (for example Boli and Thomas 1999; Djelic and Sahlin-Andersson 2006; Drori et al. 2006). For example, the FSC was established primarily at the initiative of WWF, which also exported the FSC certification model to the fisheries sector by creating the MSC. Whereas rational institutionalism sees regime formation as a functional solution to (environmental) problems in specific sectors, sociological institutionalism highlights the influence of organizational carriers who promote particular organizational recipes that are consistent with salient norm and values in the institutional environment. In a sense, the spread of the certification model could be seen as resulting from 'a solution in search of a problem' rather than a functional response to particular problems (March and Olsen 1976, quoted in Auld et al. 2007). The success of an organization is judged from its ability to adapt to popular organizational ideas and recipes, which, in turn, are legitimized by institutionalized norms and values. Consequently, a successful recipe can be expected to be consistent and aligned with salient norms and values in an organizational field. Popular organizational recipes may or may not enhance instrumental problem solving, but as long as the organization adopts those recipes, it is deemed successful by field-level audiences.

According to DiMaggio and Powell (1983), homogenization within organizational fields may occur as a result of three processes: *coercive isomorphism*, *mimetic processes* and *normative pressures*. Coercive isomorphism may result from government regulations, but it could also result from preferences for particular organizational forms from donors, charities or other funding bodies. Mimetic processes occur when a number of organizations imitate a specific organizational model that is considered particularly legitimate or successful. Organizational models may be promoted by carriers like environmental NGOs, consulting firms, management gurus or industry trade organizations. Normative pressures occur as professionals, educated in the same schools or university systems, occupy similar positions across a range of organizations, and introduce their occupational principles, norms and values in those organizations. Although all these processes may be at play in non-state institutional formation and design, I expect that mimetic processes will be particularly important as a result of the influence of a global culture comprising broad consensus on the set of appropriate organizational forms (Meyer et al. 1997) and the actions of organizational carriers like the WWF and other advocacy groups. Environmental NGOs advocate the adoption of specific organizational recipes by praising or damning industry practices, mobilizing consumers and convincing companies about the benefits of adopting those recipes.

The spread of the certification model could rewrite the rules of organizations for doing business in a more fundamental way than by restructuring incentives. It may be decided that the organization will participate in certification schemes, because certification is associated with the identity of a modern organization, because it is seen as fashionable or because it is considered to be a pre-eminent way of meeting expectations about appropriate conduct from relevant audiences. A particular certification scheme may acquire a high level of legitimacy within a sector, in the sense that participation is considered to be the right and appropriate thing to do. To use labels coined by March and Olsen (1989), organizations follow the 'logic of appropriateness' rather than the utilitarian 'logic of consequences'. Company leaders may also simply go with the flow and do what many other companies do; instead of calculating the costs and benefits of participation, it may be taken for granted that they ought to participate because other companies are participating. Yet another possibility is that there is no clear idea within the company about the consequences of their participation. Returning again to insights from regime theory, Young (1989) argues that what he calls 'a veil of uncertainty' facilitates 'institutional bargaining' in processes of international regime formation. His argument is that decision makers' uncertainty about what is in their best interest and about the future consequences of institutional arrangements enables states to form and participate in regimes. Similarly, uncertainty about the consequences of non-state governance schemes could enable agreement on institutional arrangements among the various stakeholders and facilitate producer participation.

Turning to institutional consequences, a key expectation from sociological institutionalism is that, through their participation in non-state governance schemes, producers may *internalize* norms and rules about appropriate conduct in particular roles and situations. From this perspective, social learning and internalization of norms and rules constitute the prime causal mechanism believed to connect non-state governance schemes to behavioral change. Thus actors learn and accept the norms and rules of the scheme, and then use them to guide their behavior without having to reflect upon them. March and Olsen (1989, p. 23) depict behavior as being rule-driven: 'to describe behavior as driven by rules is to see action as a matching of a situation to the demands of a position'. Instead of examining their individual goals and calculating the costs and benefits of behavioral options, then, actors try to match specific situations with the specific role called upon in this situation and the appropriate action as an occupant of that role. Rules of appropriateness are defined by political and social institutions and transmitted through socialization (March and Olsen 1989, p. 23). According to this view, institutions influence behavior, but in a different way than we would expect from an interest-based perspective.

The assumption that actors follow the logic of appropriateness and not the logic of consequences has important implications for expectations about compliance with certification standards. Whereas rewards for compliance and sanctions for noncompliance are seen as crucial from the rational institutionalism perspective, sociological institutionalists would expect companies to comply with certification rules because of learning processes, internalization of rules, and habit. As a result, the *process* of developing and learning rules becomes more important than compliance verification and enforcement. Stakeholders that consider standards and rules as legitimate are more likely to comply with them than are those that believe the standards are unfair, inequitable or unjustified. Like legal rules in international society, legitimate standards can exert an autonomous binding force and a 'compliance pull' of their own (cf. Franck 1990). According to this perspective, we expect that third-party auditing of practices is still important, but it is seen as a process whereby resource managers learn and accept rules, and then use them to guide their behavior – rather than as an instrument primarily geared toward the enforcement of compliance. Problems with cheating and free riding do not loom large if stakeholders believe that the standards have emerged from a legitimate and fair process (Breitmeier et al. 2006). In a sense, actors feel compelled to comply with standards that are considered legitimate.

In the literature on international environmental regimes, a distinction is sometimes made between the enforcement approach and the management approach to compliance (Chayes and Chayes 1995). Proponents of the management approach to compliance argue that information sharing, technical and financial assistance, implementation support, systems of implementation review and the like will be just as effective in eliciting compliance as strict enforcement of rules will be. Whereas rational institutionalists would argue that compliance with standards must be enforced, sociological institutionalists would contend that engaging producers in certification processes can elicit compliance, even in cases where noncompliance would not have negative consequences.

Summary and Analytical Implications

This discussion has pointed to various causal mechanism and pathways that could mediate between institutions and behavioral outcomes. A focus on causal mechanisms may help the analyst organize process tracing within cases and reveal the branching points and chain of events that resulted in specific outcomes (George and Bennett 2004). Two such general mechanisms have been identified. According to interest-based regime theory, institutions influence behavior by restructuring incentives; they create incentives for compliance and increase the costs of noncompliance. Standard setters offer target companies such rewards as enhanced reputation or greater market access, on the condition that the company adopts and complies with the standards. Behavioral adaptation in line with the standards is more likely to occur when actors expect the promised rewards to be greater than the costs of compliance.

According to sociological institutionalism, institutions influence perceptions about acceptable or appropriate behavior within an issue area; they create a sense of obligation to follow rules and commitments. Behavioral adaptation in line with the standards is more likely to occur when actors consider rule following as the appropriate and 'right thing to do'. These mechanisms are supplementary rather than mutually exclusive; the question is when and under what conditions each of them is likely to come into play in non-state standard-setting processes.

Beyond the expectations generated from each of the two theoretical perspectives, there is likely to be some type of interplay between the internalization of norms and rules and strategic-calculative decisions about participation in certification schemes and compliance with rules. The principles, norms, rules and governance arrangements of non-state institutions could result in a realignment or redefinition of company interests and the boundaries of acceptable and appropriate behavior. A company's decision makers may simply take for granted that they ought to participate in a certification scheme in order to obtain a societal license to operate. But they could still adapt strategically to a new reality by choosing to sign up for a less demanding scheme rather than a more stringent scheme. Another possibility is that resource managers in companies that joined certification schemes because of strategic calculative decisions learn and internalize certain environmental protection norms and rules. As a result, they will comply with those norms and rules habitually, without case-by-case deliberations about the costs and benefits of compliance (cf. Breitmeier et al. 2006, p. 155). The analysis must be sensitive to such interaction effects between the factors that influence institutional formation and effectiveness.

Evaluations about the legitimacy of certification schemes and strategic-calculative evaluations about participation are also likely to be interconnected. If the legitimacy of a certification scheme is widely questioned because of considerations about issues like equity, fairness and distributive aspects, nonparticipation can be more easily justified and can therefore be less costly for companies. Questions and concerns about the legitimacy of a particular certification scheme could also be part of a strategy for creating support for a competing scheme with different standards. The proliferation of competing schemes could, in turn, result in new struggles for achieving rulemaking legitimacy and support from a wide range of constituents (Cashore et al. 2004). Moreover, any market-based certification scheme depends on trust and moral support from relevant audiences in the marketplace. If certain salient audiences did not see a certification scheme as being legitimate and credible, there would be neither economic incentives nor normative pressures for companies to join the scheme.

ANALYTICAL APPROACH

Drawing on extant work on non-state governance and preliminary research on the cases, this section investigates when and under what conditions non-state governance schemes are likely to emerge and influence the behavior of target groups. Corresponding to the broad research questions introduced earlier, I outline factors that are likely to influence: (1) the emergence of non-state certification programs, (2) the unfolding and outcome of standard-setting processes and (3) the effectiveness of non-state certification programs.

The Emergence of Non-state Certification Schemes

The central question here is how we can analyze the formation and proliferation of non-state certification schemes in the forestry and fisheries sectors. It is fascinating in its own right to understand patterns of emergence, but it is also fundamental for evaluations of effectiveness, as producers self-select into certification schemes. As a consequence, we have to consider the possibility that these schemes attract participation only from producers that do not have to implement costly management reforms in order to comply with the standards. In this section, I propose key factors that are likely to influence the formation of certification schemes.

Inadequate public regulations

Non-state governance schemes do not exist independently of public rules and regulations. I expect that institutions for non-state governance are more likely to be formed in policy domains that are weakly regulated by public authorities than in policy domains that are strongly regulated by public authorities. The assumption is that non-state actors will be motivated to fill the governance gaps left open by public authorities, supplement weak public rules and regulations with more stringent rules or compensate for the lack of public regimes by creating private regimes (for example Cutler et al. 1999; Hall and Biersteker 2002). Since the 1980s, environmental NGOs and other stakeholders have been increasingly concerned that traditional public regulations would not offer adequate protection from deforestation and global forest degradation following irresponsible industrial logging in the tropical zone and elsewhere and that governments would fail to address these problems. The lack of a forest convention or any other legally binding agreement on forests gave environmental NGOs reasons to seek an alternative solution (Humphreys 1996; Elliott 1999). Similarly, years of overfishing that depletes fish stocks have resulted in widespread concern that governments were not willing or able to resolve the problem (Phillips et al. 2003). Hence, the expectation is that environmental NGOs and other stakeholders in the forest and fisheries sectors were motivated to create private governance schemes to compensate for what they regarded as insufficient public regulations.

NGO coalition building and producer targeting

In many policy fields, including the forest and fisheries sectors, policy networks have traditionally involved public authorities, industry associations and trade unions, with little participation from outside stakeholders such as environmental NGOs. If environmental NGOs want participation from industry associations and producers in new governance schemes, they have to challenge the exclusive rulemaking authority of these policy networks. Because such schemes ultimately depend on the collaboration of producers and other stakeholders, NGOs need to build new coalitions in favor of certification and labeling (Boström 2006a). Such coalition building is likely to occur among powerful organizational actors like environmental NGOs, industry associations and 'green' companies, domestically as well as in important export markets. The inclusion of large, powerful organizations in the certification project can be expected to be essential in order to occupy the policy field. Such organizations include companies with financial resources and specific expertise (Cutler et al. 1999), retailers with strategic positions along the market supply chain (Overdevest 2004) and social movement actors with 'moral authority' (Hall and Biersteker 2002).

Besides building coalitions with powerful organizations, NGOs and advocacy groups are likely to target producers to convince them of the benefits of participation in certification schemes and to increase the costs of nonparticipation. Whereas states have the authority to make binding rules for natural resource governance and use, non-state certification schemes depend, in principle, on voluntary producer participation. In practice, NGOs expend considerable effort in persuading or coercing producers to sign onto certification programs, using a combination of 'carrots' (for example reputational or economic benefits) and 'sticks' (for example threats of boycott campaigns). Producers can be regarded as utility-maximizing actors that need to be convinced of the economic or reputational benefits of certification and labeling (Cashore et al. 2004). Because participation in fisheries and forest certification schemes requires producers to undertake costly management changes they would otherwise not pursue, they can be expected to calculate their cost of and their gain from seeking certification. In addition to the lure of a price premium on eco-labeled products, incentives for participation can take the form of prospects for greater market access or prevention of consumer boycotts.

Transnational activist networks use a range of strategies to create demand for the certification and labeling schemes they support, including the naming and shaming of producers using practices of which they disapprove. According to Haufler (2001), the threat of advocacy group shaming is a major driver of participation in such voluntary private sector programs. However, NGOs must balance the threat of boycotts and negative campaigns against those who do not participate in coalition building and positive incentives, as their ultimate goal is to convince producers of the benefits of participation. The causal mechanism at this stage, then, is primarily the restructuring of incentives for producers that are considering whether or not to sign onto certification programs. In the long run, however, successful coalition building could result in the development of shared beliefs, values and objectives (cf. Sabatier and Jenkins-Smith 1999) and influence producers' evaluations of appropriateness.

Industry structure

The effect of NGO targeting is likely to vary among industries and countries. The size, ownership and export dependence of an operation are likely to affect its vulnerability to NGO targeting (Cashore et al. 2004). Because of their public and market exposure, large, vertically integrated forest companies involved in timber extraction, processing and sales are more likely to succumb to pressure to certify than are small forest owners. Furthermore, economies of scale render it less costly for large companies to adopt certification standards, prepare for certification processes and respond to certification audits. Dependence on environmentally concerned export markets is also likely to influence adoption choices; producers dependent on environmentally sensitive export markets are more likely to certify in the hope of avoiding transnational NGO boycotts and loss of market shares (or to increase sales) than are those who sell primarily in a domestic, more easily pacified market (Cashore et al. 2004).

Cashore et al. (2004) also found that forest companies and forest owners in a country with diffuse or nonexistent producer associations are more likely to sign onto FSC than are those in a country with strong and unified producer associations. A producer-dominated program is more likely to emerge in countries or regions with strong, well-organized associations because a strong associational system is better able to stave off NGO pressure to participate in FSC by undertaking collective and strategic industry responses.

From a commodity supply chain perspective, large companies can dictate the terms of buying and selling arrangements up and down the supply chain, resulting in demand for certification (Overdevest 2004). Horizontal modes of diffusion can result from advocacy group targeting of companies such as supermarket chains or seafood restaurants at the same level in a supply chain (Sasser et al. 2006; Auld et al. 2008). Accordingly, I expect that variation in industry structure across sectors and countries is likely to result in different responses to NGO pressure to certify.

Government support

The rise of transnational corporations, the growth of non-state actors and the diffusion of power in a globalizing economy are sometimes seen as evidence for 'the retreat of the state' in world politics (Strange 1996). More specifically related to the proliferation of institutions for non-state governance, there is talk of 'crowding out' traditional command-and-control instruments and public policies (for example Clapp 1998; Haufler 2001). Yet governments continue to regulate businesses, investors, communities, citizens and natural resource use through legal systems, property rights, taxation, planning rules and the like. States remain the primary units of the international systems through which political competition and mobilization are channeled. And most students of international relations would agree that states have legitimate rulemaking authority over and beyond non-state actors and institutions. I expect, therefore, that government support or lack of support may facilitate or hinder the proliferation of non-state certification schemes.

There are several ways that governments could influence the emergence and spread of non-state governance schemes. First, they could grant legitimacy directly to non-state governance institutions by delegating rulemaking authority (Cutler et al. 1999), or more indirectly by expressing moral support for the institutions (Boström 2003). Government support could enhance the credibility of private schemes and strengthen perceptions that pursuing the certification track is appropriate action for environmentally concerned companies. Second, governments could facilitate market acceptance of certification and labeling schemes through public procurement policies. Of course, if governments favor one certification scheme over another in public procurement policies, it would be a strong signal to firms considering various options. They could also impede the spread of non-state governance schemes by rejecting particular schemes or labels. Third, governments could facilitate private sector governance schemes at a more practical level by tendering knowledge, expert advice and financial support in the development and implementation of such initiatives (Boström 2003). I expect that these forms of government support will facilitate the emergence and proliferation of non-state certification and labeling schemes. In sum, government support is likely to influence not only producers' cost-benefit calculations concerning participation in certification schemes, but also their evaluations of legitimacy and appropriateness.

The Unfolding and Outcome of Standard-setting Processes

Non-state standards institutions constitute governing *arenas* that assemble various stakeholders, regulate their interactions and provide opportunities for learning and mutual adaptation of behavior. When we talk about an institution as a governing arena, we are interested in 'the access of actors to problems and the access of problems to decision games' (Underdal 2002, p. 24), as well as processes of learning, inclusion and adaptation (Bernstein and Cashore 2007). Governing arenas must have mechanisms for aggregating preferences into collective decisions such as decision rules and procedures. Standard setters must also decide on the type of actors that should be allowed to participate in rulemaking, and what role they should play in the governance process. In this section, I examine how organizational design is likely to influence standard-setting processes and outcomes.

Organizing the rulemaking process

The manner in which rules are developed and agreed upon can be expected to distinguish legitimate rules from those lacking legitimacy (Breitmeier et al. 2006, p. 91). I expect that the level of *inclusiveness* in standard-setting projects is particularly important for the unfolding of the process and what it produces. Inclusiveness refers to the degree to which a broad range of stakeholders is included in standard development and on the governance bodies of standards organizations. Participation by a broad range of stakeholders, representing economic, ecological and social interests, can be expected to enhance the legitimacy and credibility of a certification scheme among local communities, professional purchasers, customers and the general public. Environmental NGO participation is likely to be particularly important owing to the 'moral authority' (Hall and Biersteker 2002), specific expertise and 'symbolic capital' (Boström 2006a) of NGOs in the environmental realm.

On the other hand, producers can be expected to operate under the belief that those who must actually implement sustainability standards ought to develop or significantly influence the standards (Cashore et al. 2004). If producers feel excluded from standard development or deprived of real decision-making power, they are more likely to leave the process. There could also be a tradeoff between inclusiveness and decision-making efficiency; with an increasing number of participants involved in standard-setting processes, it may become more difficult to agree upon standards and governance procedures. The analysis must therefore be sensitive to the interacting effects of the various variables, where one variable may pull in one direction and another variable may pull in the opposite direction.

The outcome of the standard-setting process is likely to depend upon participation patterns. When industry associations and producers dominate rulemaking, their interest in keeping adoption costs as low as possible would suggest that the standard-setting process is likely to result in relatively flexible and discretionary standards. By contrast, the outcome in a multi-stakeholder arrangement is likely to be more stringent standards, because environmental NGOs tend to advocate relatively demanding environmental protection measures and base their arguments on specific knowledge claims supporting such measures. I expect that producers who participate in inclusive, multi-stakeholder, standard-development arrangements are more likely to accept stringent standards than are those who participate in industry-dominated arrangements. This acceptance can result from negotiations and social interactions with NGOs and other stakeholders that are likely to advocate stringent standards. Producers who believe that the standards have emerged from a process that is fair and equitable can also be expected to be more likely to accept stringent standards than are those who feel no sense of ownership of the outcome (cf. Franck 1990). Hence, social interactions and collaboration in standard-development processes may contribute to learning among the participants and a redefinition of their interests.

Organizing for accountability

Whereas the literature on accountability has tended to focus on accountability structures and mechanisms in democratic nation states, globalization processes and transboundary challenges have led to the emergence of governance arrangements beyond the nation state and to a renewed interest in accountability among scholars (for example Keohane 2003; Grant and Keohane 2005; Mason 2005; Newell 2005; Newell and Wheeler 2006; Boström and Garsten 2008). Accountability is a source of democratic legitimacy, not only in nation states but also in new forms of transnational governance arrangements (Dingwerth 2007). As Grant and Keohane (2005, p. 1) have stated: 'If governance above the level of the nation-state is to be legitimate in a democratic era, mechanisms for appropriate accountability need to be institutionalized'.

The creation of non-state standards organizations can be seen as an effort by civil society organizations or industry associations to institutionalize accountability mechanisms beyond the nation state. These organizations cannot simply replicate the traditional, territorial accountability structures in democratic states (cf. Grant and Keohane 2005), but they could create new tools and mechanisms that could be more effective in holding producers to account than could traditional government regulations. One way to 'organize for accountability' (Boström and Garsten 2008) is to create requirements and procedures to enhance the *answerability* of producers that adopt standards and to enhance *control* over them. Producers that adopt certification standards must consent to regular inspections of their practices and must accept the consequences of non-compliance. A certificate from a credible organization may, in turn, reassure relevant constituents and market players that a company is assuming responsibility for its conduct.

Another way to enhance accountability is to create an organizational capacity for *responsiveness* to relevant constituents. In the absence of the exclusive rulemaking authority of the state, a non-state standards organization must depend on the voluntary participation of producers and must be granted legitimate rulemaking authority by all the stakeholders it claims to represent. Just as public agencies must be responsive to the needs of clients, 'customers' and the general public, a standards organization must be responsive to a broad range of stakeholders and manage the diverse expectations generated outside the organizations (cf. Romzek and Dubnick 1987). An organizational capacity for responsiveness to clients and external constituents could be enhanced by such measures as including relevant groups in the standard-setting process, consultation with stakeholders in certification proceedings, transparent decision making, opportunities for complaints and procedures for dispute resolution.

In short, I expect that standards organizations can enhance accountability through instrumental organizational design. However, sociological institutionalists remind us that organizations may adopt certain procedures and tools because the actions of an entity must be acceptable or appropriate within a certain institutional framework. From this perspective, particular accountability recipes could be seen as *rationalized myths* (Meyer and Rowan 1977) that spread rapidly in both private and public organizations. It is interesting to explore, then, not only *how* standard setters organize to enhance accountability, but also *why* they adopt certain accountability tools and *what* effects those tools have on producer behavior.

The Effectiveness of Non-state Certification Schemes

Beyond serving as governing arenas, some non-state governance schemes also qualify as organizational *actors* with the capacity to direct the behavior of organizational members. From the governance perspective adopted in this study, the effectiveness of a certification scheme is primarily a matter of its influence on the management and use of natural resources. In the following, I examine factors that can be expected to influence the effectiveness of certification schemes.

Producer participation

In principle, participation in certification schemes is voluntary. The more producers participate in a certification scheme, the more likely it is that it will change widespread producer practices. Without a critical mass of producers, a voluntary scheme is unlikely to change producer practices in ways that lead to improvements in the biophysical environment. Accordingly, the adoption of a certification scheme can be expected to influence its problem-solving effectiveness. In addition to considering measures such as forest area certified and the proportion of certified to uncertified forests, however, it is critical to examine *patterns of standards adoption*. Because participation in certification schemes is voluntary, it is possible that only producers who face relatively low costs of standards adoption choose to participate. If producers who face substantial compliance costs were to systematically opt out of certification schemes, the net impact of certification would be low (Auld et al. 2008). Patterns of adoption can be expected to be related to the stringency of the standards, which I turn to next.

Stringency of the standards

The stringency of certification standards is likely to be critical for the environmental problem-solving capacity of a certification scheme. By stringent standards, I mean that they are prescriptive and comprehensive, requiring forest companies to limit harvesting near rivers and protected areas, for example, and fishing vessels to use particular fishing gear and methods. As a point of departure, we may expect that the more stringent the environmental standards, the greater the likelihood that they will change forestry and fishing practices in ways that lead to environmental amelioration. Stringent certification standards may compel producers to go beyond compliance with public rules and undertake costly reforms that they otherwise would not pursue. A standard requiring large forest set-aside areas, for example, would preserve larger high conservation-valued forest areas than would a less stringent standard. Although we should not expect a linear relationship between standard stringency and impact on the biophysical environment, stringent standards are likely to increase the ameliorative effects of a certification scheme on harmful producer practices.

On the other hand, stringent standards could also have negative effects on the overall effectiveness of certification schemes. First, there could be an inverse relationship between stringency and the adoption of schemes by producers, because producers do not necessarily accept schemes with demanding and intrusive standards (Cashore et al. 2004; Gulbrandsen 2004). Unless participation in schemes with stringent standards is rewarded in some way, we may expect that the more stringent the certification standards, the less likely it is that a wide range of producers is willing to participate voluntarily. More specifically, we may expect that only environmental frontrunners, which could adopt standards without having to undertake costly management reforms, would find it attractive to participate in a scheme with highly demanding and prescriptive standards. Enthusiasm for stringent certification schemes among environmental laggards, where the need for changing management practices is more urgent, can be expected to be low. Consequently, stringent standards could reduce a scheme's ability to attract participation from producers, and without a critical mass of producers, a scheme is unlikely to change widespread producer practices in ways that lead to industry-wide environmental improvements.

Second, there could be an inverse relationship between compliance and the stringency of the standards, because producers do not necessarily have the *capacity* to implement and comply with highly demanding standards. Even if producers would like to change management practices and comply with stringent standards, they may fail to do so simply because the standards are too demanding. As a result, the level of noncompliance can be expected to be higher in schemes with stringent standards than in schemes with lenient standards.

Third, it is critical to recognize that standards are not neutral; the first movers who create the rules can tailor them to match their technical and operational capacities, resulting in higher switching costs for late movers (Mattli and Büthe 2003; Auld et al. 2008). Accordingly, standards secure advantages for certain producers and disadvantages for others, and stringent standards may be tailored to enhance the competitive advantages of first movers.

To summarize, I do not expect to find a simple relationship between stringency and effectiveness. On the one hand, stringent standards may direct producers' behavior and force them to undertake reforms they otherwise would not pursue; on the other hand, there may be an inverse relationship between stringency and producer participation and, likewise, between stringency and compliance. The empirical analysis will have to shed light on when and under what conditions stringent standards are likely to result in changes in problem-relevant behavior among a wide range of producers – not merely among environmental frontrunners.

System operation

A certification scheme may be performance based (focusing on outcome), management-system based (focusing on process) or based on some combination of the two. In a performance-based scheme, compliance with standards must be verified in on-the-ground audits. When performance-based standards are assessed, the forest or fishery itself is evaluated. For example, a certifier may inspect a forestry organization to ascertain if it has set aside primary forests of a certain size or a certifier may inspect a fishing vessel to see if appropriate fishing gear and practices are in use.

By contrast, a management-system-based scheme does not dictate compliance with any specific performance level before issuing a certificate, but requires that continual process improvements be demonstrated in audits. When system-based standards are audited, it is not the forest or fishery that is assessed, but the forest or fishery management system. For example, a certifier may inspect an organization to see if it has implemented adequate management plans, internal monitoring systems and reporting procedures. An undertaking certified in accordance with system-based standards is usually required to have an environmental policy and goals in place, but can generally decide the environmental performance level it aims for. The management-system-based approach is sometimes perceived as being more dynamic than performance-based systems because of the requirement for continuous improvement rather than clearly defined and in some cases, static criteria.

On the other hand, management-system-based certification has been criticized for providing little incentive for firms to go beyond the minimum requirement of meeting domestic laws and regulations (for example Clapp 1998; Krut and Gleckman 1998). Moreover, compliance with these standards can, in principle, be verified without a visit to the forest or the fishery. Because performance-based schemes require compliance with substantive on-the-ground standards, we may expect that they are more likely to modify forestry and fisheries practices in ways that lead to less environmental deterioration than will management-system-based schemes.

Compliance with the standards, rules and policies of non-state standards organizations may be based on first-party verification (self-inspection), second-party verification (inspection by an industry or trade association) or third-party verification (inspection by an independent auditor). It is generally assumed that third-party audits of management practices and performance would constitute a stronger push toward compliance than would first-party or second-party inspections.

The assumption that third-party auditing will result in improvements is essential to all certification schemes, but auditing practices are likely to vary among schemes. Regular third-party audits by independent certification bodies could enhance compliance with standards and continuous performance improvements in certified companies. On the other hand, if auditing practices are lenient or based on highly discretionary standards, obligations to report performance and verify compliance with standards could merely become ceremonial rituals aimed at justifying the business-as-usual situation (Power 1997). In this view, prescriptions about consultation with stakeholders in standard-setting processes, accreditation of independent certifiers, third-party auditing and the like are rationalized myths that spread rapidly in both public and private organizations. Thus the analysis needs to explore the *behavioral* consequences of the certification and auditing process, to which I turn below.

Consequences of noncompliance

From the perspective of rational institutionalism, the key to achieving high levels of compliance lies in the role of enforcement; to be an effective certification scheme, the consequences of noncompliance with certification standards must be tangible enough to increase the costs of noncompliance and thus deter violations of rules (Breitmeier et al. 2006). In the case of noncompliance, the certification body would normally issue corrective action requirements and give the producer sufficient time to improve operations, but the certification body may also suspend the certificate if the producer fails to correct serious breaches of standards. In some schemes, the consequences of failing to comply could also lead to expulsion from an industry association.

As a point of departure, I expect that the more serious the consequences of noncompliance, the greater the potential to change behavior and the more effective the certification scheme will be. Failure to comply with certification standards may not only result in the loss of a certificate, but could result in loss of reputation and trust. If, on the other hand, there are no tangible consequences of noncompliance, certified producers would, from the rationalist perspective, have no incentive to change their practices in order to comply with demanding standards. Among the causal mechanisms believed to mediate between a certification scheme and improved environmental practices in forestry and fisheries, auditing largely involves the restructuring of incentives.

As a result of repeated interaction between producers, certification bodies and other stakeholders, however, producers may begin to follow rules without considering if rule following is compatible with their material self-interest. This behavior would be consistent with what we would expect from actors motivated by the logic of appropriateness, and suggests that compliance verification becomes less important over time than does maintaining a dialogue with a wide range of stakeholders in order to meet their expectations and needs. Through their participation in certification schemes, producers, as stewards of natural resources, may internalize norms and rules about appropriate conduct, and incorporate compliant behavior into standard operating procedures. Resource managers may comply with rules because it is the right thing to do or they may comply habitually, without case-by-case deliberations about the costs and benefits of compliance.

Engaging actors in such complex social networks as multi-stakeholder standard development and certification processes can produce positive results over and above the development of compliance mechanisms (cf. Reinecke 1998). As noted by Breitmeier et al. (2006, p. 155), the imposition of penalties or the provision of rewards may prove effective in eliciting compliance at the margin, but 'even well-endowed public authorities would run into trouble right away unless most subjects complied with the relevant rules and commitments most of the time without regard to the impact of punishments and rewards'. I expect, then, that engaging producers in standard-setting and certification processes can elicit compliance with rules and behavioral change, regardless of the character of the compliance system. Indeed, rational models of compliance-enforcement systems, if taken at face value, can have misleading practical implications and may even undermine the trust they are meant to build (Hasenclever et al. 1997, p. 170). Expanded monitoring and auditing and rigorous compliance verification systems can lead to an ever-growing demand for more monitoring and auditing (Power 1997). The result could be that auditing becomes an end in itself rather than a means to change problem-relevant behavior.

METHODOLOGY AND MATERIAL

This book focuses on certification programs in the forest and fisheries sector as these programs are among the most advanced cases of non-state rulemaking and governance. The research design is the comparative case-study approach, but I do not seek to achieve controlled comparisons in the sense prescribed by Lijphart (1971, p. 683). In his definition, scientific explanation consists of two basic elements: (1) to establish general empirical relationships among two or more variables and (2) to control for all other variables that represent rival explanations. This logic of inference is also central to the seminal work of King et al. (1994) on qualitative research. Other comparativists have noted that whenever social scientists examine large-scale political changes, they find that it is usually *combinations* of conditions that produce change (Ragin 1987, p. 24). This is not the same as arguing that change results from the influence of many variables, as in the statement 'gender, education and tenure affect income'. The latter argument asserts that all variables have independent effects on income. In a case of complex causality, it is the intersection of a set of conditions in time and space that produces change. If one of these conditions is missing, the change will not occur (Ragin 1987, p. 24).

Most of the national-level events that comparativists study demonstrate a great deal of causal complexity. An outcome of interest to social scientists rarely has a single cause. It is well known, for example, that the contribution of democratic rule to political stability depends on a number of other factors. Some developing countries are thought to be stable because they are democratic; others are thought to have failed because political instability increased after the adoption of democratic procedures (Ragin 1987, p. 24). To complicate the situation further, several variables or combinations of conditions can produce the same changes. This phenomenon is commonly referred to as equifinality. Moreover, a specific cause may have opposite effects, depending on context. Changes in living conditions have been shown to increase or decrease the probability of strikes, for example, depending on other social and political factors (Ragin 1987, p. 27).

Rather that seeking to achieve controlled comparisons in the strict sense, the purpose of comparing forest and fisheries certification programs is to examine similarities and differences in the conditions that influence the emergence and effectiveness of these programs. Comparing certification initiatives in the forest and fisheries sector seemed to be interesting for several reasons.

First, because fisheries certification was modeled on forest certification, it is possible to examine the transformation of the certification model as it is exported from one sector to another. This process enables an investigation of the ways certain organizational forms come to exist and travel across sectors.

Second, given the differences in the way forests and fisheries are governed, it is possible to study how the divergent roles of certification programs in those sectors affects program design and rulemaking. Whereas forests are typically managed by private owners or companies with logging concessions from governments, coastal and open-ocean fish stocks are common-pool resources that are often managed through international and regional fisheries agreements.

Third, the different nature of forest and fisheries resources enables a comparison of the challenges to credible auditing of compliance with standards. Auditors can usually observe the direct effects of forestry operations in on-the-ground inspections. By contrast, the absence of easily observable effects of noncompliance, the nonselective nature of many fishery harvest techniques and the multiple access rights to shared fish resources render auditing more complicated in the fisheries sector.

Fourth, the comparative case study design generally enables the possibility of direct replication. Empirical findings and analytical conclusions arising independently from two different cases will be more powerful than those coming from a single case (Yin 2003, p. 53).

Given the problems achieving controlled comparisons, the question is how to increase the validity of the findings of this study. Because of the limited number of cases under investigation, this study of forest and fisheries certification allows within-case process tracing. Process tracing is a procedure by which the researcher identifies causal chains of events and path dependencies that resulted in particular outcomes (George and Bennett 2004). Unlike controlled comparison, the method does not attempt to achieve conditions similar to those of an experiment.

Process tracing is often guided by the specification of one or several causal mechanisms. The specification of theoretically based causal mechanisms is valuable for understanding the relationship between independent variables and particular outcomes. A focus on causal mechanisms can help to organize process tracing within cases. In particular, it is possible to test if the chain of events and path dependencies one uncovers is consistent with one's expectations, based on a particular causal mechanism. In this way, process tracing can be explained as 'an operational procedure for attempting to identify and verify the observable within-case implications of causal mechanisms' (George and Bennett 2004, p. 138).

The book draws on primary research on certification initiatives in the forest and fisheries sectors conducted over several years. Guided by the analytical framework, the case studies follow a similar structure in order to facilitate comparisons. Moreover, the process tracing and reconstruction of chains of events are based on the same type of information. Process tracing was made possible by investigating journal, magazine and newspaper articles and electronic information. Semi-structured interviews with representatives of MSC's secretariat in London, UK, have supplemented the written sources on fisheries certification. Regarding forest certification, semi-structured interviews with representatives of certification programs, environmental NGOs, forest owner associations, industry associations and government agencies in Sweden and Norway have complemented the written sources. Secondary literature on forest certification in other countries has supplemented the interview material from Sweden and Norway. As noted in Chapter 1, I found the work of Cashore et al. (2004) on forest certification in Europe and North America to be particularly helpful in guiding the analysis of the emergence of certification programs. My collaborative work with Auld and McDermott

(Auld et al. 2008) has informed the analysis of the effectiveness of those programs.

In summary, the analysis is based on a comparative study of certification initiatives in the forest and fisheries sectors. The comparison of forest and fisheries certification provides a fuller and richer account of the origins and effects of non-state certification programs than single case studies do. In addition, it is a multi-level study, including the global and national levels for both sectors. The in-depth study of forest certification in Sweden and Norway provides a deeper understanding of the factors that influence institutional emergence and effectiveness. This combination of comparisons across governance levels and sectors serves to strengthen the reliability and validity of the findings.

3. The emergence of forest certification

Forest certification emerged in response to increased international attention to global forest degradation and prolonged efforts within intergovernmental bodies to develop a legally binding agreement on forests. The lack of a forest convention or any other legally binding agreement on forests gave NGOs and other organizations concerned about forest destruction reasons to seek an alternative solution.

This chapter examines the creation and evolution of forest certification schemes. As a foundation for assessing these schemes, it begins with a review of intergovernmental efforts to address forest degradation and deforestation. The second section examines the creation of FSC by a broad coalition of stakeholders and the emergence of producer-backed programs with more discretionary and flexible standards. The third section begins with a comparison of certification standards across various programs, demonstrating that the poorer performers – the producer-backed programs – have increased the stringency of their standards over time. A comparison of auditing procedures reveals a similar pattern: the poorest performers have adopted several conventions to improve their auditing processes. An explanation is then sought for the increased stringency in the standards and auditing procedures of producer-backed programs. The fourth section demonstrates that although producer-backed programs have mimicked some of FSC's governance arrangements, different approaches to stakeholder involvement indicate that the distinction between a multi-stakeholder and a producer-dominated governance model still applies. The conclusion reflects upon whether the changes in producer-backed programs reflect a real commitment to broaden and deepen the social and environmental leverage of certification or if they are primarily symbolic window-dressing.

INTERGOVERNMENTAL FOREST POLICY PROCESSES

Despite increasing concern over global forest degradation and deforestation in the tropics and elsewhere, states have failed to agree upon a legally binding global agreement for the protection and sustainable use of forests. Intergovernmental efforts to address tropical deforestation initially focused on the International Tropical Timber Organization (ITTO), created in 1986 to implement the first International Tropical Timber Agreement (ITTA 1983) – the first commodity agreement to include a conservation component. Environmental NGOs were soon disappointed over the failure of ITTO to deal effectively with deforestation and the serious environmental problems facing tropical forest management. This lack of action led environmental groups and several developed countries to advocate the idea of a legally binding global forest convention. During the preparatory process for the 1992 United Nations Conference on Environment and Development (UNCED) in Rio de Janeiro, nine proposals for a global forest agreement were tabled. Forest-rich tropical countries were adamant that a forest convention infringed on their sovereignty and this resulted in the failure of all the proposals (Humphreys 1996). At UNCED, developing and developed countries agreed instead on the Forest Principles, which are general guidelines for the management of forests relating to economic, environmental and developmental concerns. Being a set of nonbinding principles that does not clarify how conservation and utilization of forests should be balanced, the agreement is legally and politically weak.

To clarify and expand upon the Forest Principles, intergovernmental collaboration took place under the auspices of the UN Commission on Sustainable Development (CSD) in the Intergovernmental Panel on Forests (IPF) from 1995 to 1997 and in its successor, the Intergovernmental Forum on Forests (IFF) from 1997 to 2000. The most controversial issue at the IPF and IFF was whether or not to seek agreement on a global forest convention. The IPF and IFF produced a number of forest policy recommendations, but states could not agree on the need to begin negotiations on a forest convention. The IFF Proposals for Action (2000), endorsed by the UN, recommended the establishment of a United Nations Forum on Forests (UNFF) to continue cooperation on international forest policy. Unlike its predecessors, UNFF has universal membership and reports directly to the UN Economic and Social Council (ECOSOC). Therefore it has a higher profile in the UN system than either of its predecessors, but it is not equipped with facilitative or enforcement mechanisms that could enhance the follow-up of international forest policy recommendations.

It was the belief within WWF and other NGOs that by circumventing intergovernmental forest policy negotiations, forest certification would offer an alternative, fast-track route to sustainable forest management and forest protection around the world. Rather than operating alone and in isolation from intergovernmental processes, however, forest certification programs are embedded in a broader international policy domain. Understanding the origins and effects of certification programs requires that this context be taken into account. Before examining the emergence of forest certification programs, it can be useful to review some of the achievements and limitations of intergovernmental efforts to promote forest protection and sustainable forest management. Although by no means an exhaustive account of forest-related agreements and recommendations, such a review also helps to explain what certification does – or could do – for forest protection and forest management in the absence of a legally binding convention.

An Integrated Ecosystem Approach and Protected Areas

Because most of the world's remaining terrestrial biodiversity is found in forests (UNEP 1995, p. 749), many of the provisions under the Convention on Biological Diversity (CBD) have – or should have – implications for forest management and conservation policies. The CBD has established the principle of *in situ* conservation – 'the protection of ecosystems, natural habitats and the maintenance of viable populations of species in natural surroundings' (UN 1992a, article 8d). A number of other multilateral environmental agreements also promote the ecosystem approach and the principle of protected areas (see Humphreys 2006). Likewise, several soft law agreements on forests, including the Forest Principles and the IPF and IFF Proposals for Action, refer to an ecosystem approach and protected forest areas. The main shortcoming of the ecosystem approach in forest-related agreements is, however, that it is only a normative principle – it has no regulatory 'bite.'

Similarly, there are neither commitments nor guidance on the size and nature of protected areas. In 1998, the parties to the CBD agreed upon a work plan for forest biodiversity, aimed at integrating the conservation and sustainable use of biodiversity in national forest policies. This work resulted in an expanded program of work on forest biological diversity, adopted by the parties to the CBD in 2002 (UNEP 2002). To date, it has been closer to a program of research and an exchange of information than to a development of forest policy commitments. By failing to clarify how conservation and use of forests should be balanced, the CBD and other multilateral agreements, as well as the UN forest process, do little in reality to reverse deforestation and limit the commercial utilization of forest resources. Unique and valuable forest ecosystems, as well as endangered and vulnerable species and their forest habitats, still lack adequate international legal protection.

Participation, Traditional Forest-related Knowledge and Equitable Sharing

Another set of commitments and recommendations in international agreements pertaining to forestry is related to the participation of various stakeholders, the recognition of traditional forest-related knowledge and the equitable sharing of the benefits accruing from the use of forest resources. The Forest Principles urge governments to encourage the participation of economic, social and ecological interests in national forest policies and to support the identity, culture and rights of indigenous people and local communities (UN 1992b, principles 2d and 5a). Similarly, the principle of participation is explicitly mentioned in the IPF and IFF proposals, Agenda 21 and the CBD. Equally important to indigenous peoples and local communities is the CBD principle of equitable sharing of benefits from the use of genetic resources. According to Rosendal, the discussion on traditional forest-related knowledge in the IFF 'largely paralleled those in the negotiations leading up to the [CBD] eight years previously in Rio' (Rosendal 2001, p. 453). The result was that most of the IFF language reaffirms the provisions of the CBD on this issue, and explicit references are made to CBD Articles 15, 16 and 19. Economic forest interests prevail throughout the nonbinding forest agreements, however, and there are few restrictions on investments in forest areas (Humphreys 2003). Thus, the forest-related agreements arguably favor those with an interest in the commercial utilization of forests.

Criteria and Indicators of Sustainable Forest Management

The primary purpose of criteria and indicators is to report on forest conditions at the national level, allowing governments and policy makers to share information and use comparable parameters to describe the state of forests (Rametsteiner and Simula 2003, p. 91). States have agreed on nine sets of regional criteria and indicators for sustainable forest management, including the ITTO criteria and indicators (adopted by 28 tropical timber-producing countries in 1992), the Pan-European criteria and indicators (adopted by 36 countries in 1993) and the Montreal-process criteria and indicators for non-European temperate and boreal forests (adopted by 12 countries in 1995). About 150 countries have participated in one or more of the nine regional criteria and indicator processes (FAO 2001, p. 54) Through the work in the regional processes and in the IPF and IFF, a consensus has emerged on criteria and indicators adapted to different forest types and regions of the world. Although this common understanding of what constitutes sustainable forest management is valuable, it must be remembered that sets of criteria and indicators contain no targets, timetables or performance requirements (Rametsteiner and Simula 2003, p. 91). A set of criteria and indicators basically constitutes a tool for information sharing – not prescriptive standards for well-managed forests.

Illegal Logging and Associated Trade in Forest Products

Although the principles and rules of the World Trade Organization (WTO) have precluded all governmental efforts to ban trade in illegally logged timber, illegal logging featured as a key element in the G8 Action Programme on Forests (1998–2002), and led to a series of regional Forest Law Enforcement and Governance (FLEG) conferences. The geographical regions covered by FLEG processes were East Asia and the Pacific (launched in Bali, Indonesia in September 2001), Africa (launched in Yaoundé, Cameroon in October 2003) and Europe and North Asia (launched in St Petersburg, Russia in October 2005). These regional processes to counter illegal logging were initiated by the US State Department, which considered UN institutions like UNFF too slow to deal with the complex issues involved (Humphreys 2006). Primarily as a result of the FLEG processes, however, institutions such as UNFF and ITTO are beginning to address illegal logging. In January 2006, states agreed upon the third International Tropical Timber Agreement (ITTA). Whereas the first ITTA of 1983 did not mention illegal logging, and the second ITTA of 1994 acknowledged only the 'undocumented trade' in forest products, the problem of illegal logging is explicitly recognized in the third ITTA. As the first multilateral agreement to address illegal logging, agreement on the third ITTA was a significant event, although Brazil and some other producer countries have strong reservations about this part of the agreement.

To complement the FLEG processes, the EU committed in February 2002 to developing an action plan to counter illegal logging. As a major timber importer, the EU's aim was to develop supply-side measures to curtail the trade of illegally logged timber to the EU while providing assistance to producer countries to support such measures (Humphreys 2006, pp. 156–9). This focus on trade led the EU to extend the FLEG acronym when developing what became known as the Forest Law Enforcement, Governance and Trade (FLEGT) action plan. The action plan was approved in Council Conclusions in October 2003. A key element in the plan is the development of voluntary partnership agreements between producer countries and the EU on timber licensing. Producer countries that enter into such agreements with the EU commit to exporting to the EU only legally logged timber. Other measures include member states' adoption of public procurement policies to purchase timber from legal sources;[1] the promotion of private sector initiatives, including support for high standards in codes of conduct, transparency in the private sector and independent monitoring; and the exercise of environmental and social due diligence by export credit agencies and financial institutions that fund logging in producer countries (European Commission 2003).

The principles and rules of the WTO have influenced the design of the FLEGT action plan and licensing scheme (Humphreys 2006, p. 157). Rather than insisting that the scheme apply to all timber-producing countries, which would likely have encountered a challenge at the WTO, the EU decided to implement the scheme through bilateral, voluntary partnership agreements. Humphreys (2006, p. 157) explains that because the scheme is voluntary, 'illegal loggers who successfully evade the authorities in the producer country can (then) circumvent the licensing scheme by exporting timber to a country with no VPA [voluntary partnership agreement] for onward shipment to the EU'. Although the scheme is compulsory for any producer country that enters into a voluntary partnership agreement with the EU, illegally logged timber can continue to enter the EU from producer countries that have no such agreement. This loophole, which is a consequence of the limits WTO law places on measures to curtail the trade of illegally logged timber, represents a major weakness of FLEGT (Humphreys 2006, pp. 166–7).

Carbon Sequestration

Forests are recognized as important carbon reservoirs – or sinks. Reducing emissions from deforestation and forest degradation (REDD) has captured international attention as a potentially cost-effective climate change mitigation option (Stern 2006). Forest loss, primarily tropical deforestation and forest degradation, accounts for almost 20 percent of global greenhouse gas (GHG) emissions. The ability of a forest to capture and sequester CO_2 is acknowledged under the UN Framework Convention on Climate Change (UNFCCC 1992) and the Kyoto Protocol (1997). Annex 1 parties to the Kyoto Protocol may use *afforestation* (planting of new forests) and *reforestation* (planting of forests on lands that historically contained forests), measured as verifiable changes in carbon stocks since 1990, to meet their emission targets (UN 1997, Article 3(3)). Conversely, deforestation in Annex 1 countries since 1990 may have a negative impact on their balance of carbon stocks. The Clean Development Mechanism (CDM) of the Kyoto Protocol allows emission credits for afforestation and reforestation projects during the first commitment period (2008–12), but not for projects related to forest protection and the sustainable management of existing forests.

REDD projects are not recognized under the CDM during the first commitment period, but there is a growing international consensus that a post-2012 UN climate-change treaty must include incentives to reduce GHG emissions from forests. At COP 13, held in Bali in December 2007, the parties to the UN Framework Convention on Climate Change (UNFCCC) agreed under the Bali Action Plan to consider: 'Policy approaches and positive incentives on issues relating to reducing emissions from deforestation and forest degradation in developing countries; and the role of conservation, sustainable management of forest and enhancement of forest carbon stocks in developing countries' (UNFCCC 2007, Decision 1b [iii]).

Sequestering may create stronger incentives for forest protection than do any of the soft law forest policy recommendations, although it could also stimulate uniform 'carbon plantations'. Viewing forests primarily as sinks – or carbon reservoirs – is different from appreciating their value in the full range of plant and animal species they accommodate. The planting of fast-growing monocultures of softwood would be the most cost-efficient way to lock up CO_2 and thereby claim emission credits, but this is certainly not a measure that could ensure species diversity. Carbon sequestration may even result in actions to *replace* natural grown forests with plantations, something that will almost certainly result in a loss of forest biodiversity. Similarly, substitution of biomass energy for fossil fuels – an implicit incentive of the Kyoto Protocol – could result in more intensive forestry at the expense of biodiversity conservation. There is also scientific uncertainty on the role of forests as long-term carbon sinks.

National Forest Programs

National forest programs (NFPs) are recognized as a means of implementing internationally agreed-upon principles of sustainable forestry. Since UNCED in 1992, the UN Food and Agriculture Organization (FAO) has assumed a leading role in promoting NFPs, and there is now consensus that such programs should integrate forests in holistic land-use plans (Humphreys 1999, p. 252). States must report on the performance of forest policy commitments, and ECOSOC has tasked the UNFF with monitoring the progress of implementation. The problem is, however, that whereas the commitments and recommendations of international agreements pertaining to forestry are important, most of them are not legally binding and – being primarily a collection of normative principles without rules, targets or timetables – it is difficult to ascertain the degree of implementation. And because the 'forest regime' is fragmented and based on a number of agreements rather than a single convention or protocol, the UNFF can, for the most part, monitor progress only in the implementation of the recommendations that have been agreed upon under the cluster of nonbinding forest agreements. Nor are there effective enforcement or facilitative mechanisms in the UNFF to promote implementation of intergovernmental forest policy proposals.

Summary

In summary, there are a number of shortcomings in intergovernmental forest policy processes and forest-related agreements. First, there is the failure to give environmental and social stakeholders a voice in intergovernmental forest policy processes. We saw that the interests of forest owners tend to prevail throughout state-centered processes and in forest-related agreements. Second, there is the lack of strong environmental and social commitments in international forest-related agreements. States have failed to agree upon substantial measures for protecting forests, promoting a holistic ecosystem approach to forestry and ensuring the interests of indigenous peoples and local communities. A third weakness is the lack of effective control and compliance mechanisms. Although measures have been adopted to promote reporting and review of the implementation of international forest policy recommendations, no effective instruments exist for enforcing or facilitating compliance. Fourth, there are no multilaterally agreed-upon rules for trade in products sourced from well-managed forests. Given these shortcomings in intergovernmental forest policies, forest degradation and tropical deforestation continues, largely unabated, in spite of intergovernmental efforts to address the problems.

THE FORMATION OF FOREST CERTIFICATION SCHEMES

Certification emerged alongside intergovernmental forest policy processes, but interest in certification as an instrument of environmental and social governance grew after the failure to agree on a forest convention. This section reviews the creation of FSC by a broad coalition of stakeholders, its principles and governance arrangements, the emergence of FSC-competitor programs created by forest landowner or industry associations and efforts to achieve mutual recognition of the various certification programs.

The Creation of the Forest Stewardship Council

In the early 1980s, NGOs introduced boycotts, demonstrations and public shaming campaigns to target logging companies and large retailers that were sourcing tropical wood products. They soon realized that these measures alone would not resolve the problems with rainforest destruction; buyers needed a guide to identify wood sourced from well-managed forests. From 1988 onwards, Friends of the Earth UK distributed a 'good wood' buyers' guide and distributed 'seals of approval' to manufacturers and retailers (Counsell and Loraas 2002). The system was based upon a code of conduct and selected forest audits, but Friends of the Earth did not have the capacity to monitor and enforce compliance and track wood products sourced from approved operations. Long and complex wood commodity chains made it difficult to verify a product's environmental merits; some large retailers have hundreds of wood suppliers originating from thousands of forests in a number of countries. Whereas some retailers began selling wood approved by Friends of the Earth, others made unverified claims about the sustainability of the wood products they were sourcing (Klooster 2005, p. 406). Meanwhile, in the USA, the Rainforest Alliance created the SmartWood Program to audit domestic and tropical forestry practices. Launched in 1989, the SmartWood Program became the world's first forest certification scheme, followed two years later by the California-based Scientific Certification Systems forest conservation program (Cashore et al. 2004, p. 99). Both organizations now operate as FSC-accredited third-party certifiers.

Environmental NGOs also lobbied intergovernmental forest negotiations to create support for a global certification system. By the late 1980s, NGOs had become frustrated with the failure of the ITTO to promote tropical forest protection. In 1989, a Friends of the Earth proposal, tabled by the UK delegation to ITTO's Permanent Committee on Economic Information and Market Intelligence, recommended the development of a government-sanctioned global system for the certification and labeling of tropical timber. The proposal failed largely because of opposition from industry representatives and their governmental allies from tropical countries. The refusal of ITTO even to consider developing a certification scheme convinced WWF that such a system had to be developed by private initiative (Humphreys 1996, pp. 72–5). WWF's conviction gained strength during the preparatory process for the 1992 Earth Summit in Rio de Janeiro, given the lack of support that was shown for the aspiration of negotiating a legally binding forest convention. As Francis Sullivan of WWF-UK argued, 'You can't just sit back and wait for governments to agree, because this could take forever' (quoted in Murphy and Bendell 1997, p. 106).

Having witnessed the dwindling of consumer movements in the early 1990s, WWF and other environmental NGOs realized that they needed participation from retailers and other commercial interests to create such a certification program. As Counsell and Loraas (2002, p. 12) explained, 'The early promoters and founders of the FSC thus found themselves needing to enlist the support of commercial interests as much as the commercial interests needed certification as a way to placate consumer protesters'. Beginning with a meeting in San Francisco in 1991, they formed an informal certification working group with participation from retailers, forest company officials, social groups and government officials. According to Timothy Synnott, FSC's first executive director, 'over the next year, most of the activities that led to the founding of FSC were associated with this group or its members' (Synnott 2005, p. 13). Initially, there was little clarity about the role of the new organization under development, already referred to as FSC, and its relationship to professional certification bodies. It eventually became clear that unlike the existing programs, SmartWood and Scientific Certification Systems, FSC would not certify forests itself, but would set standards for the certification of forest operations and for the accreditation of third-party certifiers. A 1992 meeting of the group in Washington, DC, established an interim FSC Board of Directors and discussed the draft of a charter, standards and an operational manual for a forest certification system. At this meeting, it was also agreed that the group would put a process of widespread consultation into motion to gauge support for establishing FSC and preparing a founding assembly (Synnott 2005). From then on, FSC had a de facto operational existence, and because certifiers were already operating, the first FSC certificates – a forest management certificate in Mexico and a chain of custody certificate in the USA – were issued.

In 1993, 130 participants from 26 countries met in Toronto, Canada, to establish FSC formally. It was officially founded to promote 'environmentally appropriate, socially beneficial and economically viable' forest management. Participants included representatives from environmental NGOs, social groups, retailers, manufacturers, forest companies and professional certification bodies. WWF was the architect behind FSC, but the organization collaborated closely with other, more radical NGOs in creating support for the program. While WWF developed FSC and created buyers' groups to support it, the more radical NGOs like Greenpeace and Friends of the Earth generated supply-chain pressure for certification by targeting highly profiled companies, including the world's largest do-it-yourself retailer, the US-based Home Depot, and another large do-it-yourself retailer, the UK-based B&Q. This combination of WWF's outreach to the forest industry and the threat of shaming and boycotting campaigns from more radical NGOs was crucial in fostering support for FSC in the forest product manufacturing and trade sectors (McNichol 2002; Klooster 2005). The market campaigns from environmental NGOs created a demand for FSC-certified wood even before the program was up and running.

The FSC Principles and Governance Arrangements

Two years after its official inception, FSC was legally registered as a non-profit organization in Oaxaca, Mexico. It is constituted as a membership organization in which ultimate authority rests with its membership. The highest decision-making body in FSC is the General Assembly, which has a tripartite decision-making structure consisting of social, environmental and economic chambers. Each chamber holds one-third of the votes in the General Assembly, regardless of the size of its membership. Both organizations and individuals committed to FSC's goals can become members, but individual members are allowed to hold only 10 percent of the voting rights within each chamber (FSC 2002a). The tripartite chamber structure was designed to ensure that no specific interests are allowed to dominate rulemaking in the scheme, and, in particular, to avoid industry domination, which many viewed as a problem with intergovernmental processes (Humphreys 1996). In addition, each chamber is subdivided into a Northern (developed countries) and Southern (developing countries) division, with voting parity. The purpose of this arrangement was to empower developing country stakeholders, who often become marginalized in state-centered processes.

Membership in the economic chamber is open to companies that have committed to FSC certification, accreditation or retailing – forest companies, forest owners, processors, retailers, certification bodies and others with a financial interest in forestry. Membership in the social and environmental chambers is open to not-for-profit NGOs and to individuals who have demonstrated commitment to FSC's goals. Whereas most members in the environmental chamber are environmental NGOs or individual members, the social chamber comprises a more diverse group of members, including representatives from local communities, trade unions, indigenous peoples' groups and social NGOs, as well as individual members. Because FSC was created in opposition to intergovernmental forest policy processes, governments were not allowed to vote or participate in any of the chambers (Gulbrandsen 2004), although, as of 2002, government-owned and controlled companies may apply for membership in the economic chamber (Meidinger 2006). Table 3.1 provides a general overview of FSC membership as of December 2009, showing that only 18 percent of FSC members belonged to the social chamber, compared to 41 percent each for the economic and environmental chambers. Whereas northern members have tended numerically to dominate FSC, the table

Table 3.1 Overview of FSC members, as of December 2009

	Northern members		Southern members		Total	
		(Percent total)		(Percent total)		(Percent total)
Economic chamber	182	(22)	154	(19)	336	(41)
Environmental chamber	132	(16)	210	(25)	342	(41)
Social chamber	57	(7)	94	(11)	151	(18)
Total	371	(45)	458	(55)	829	(100)

Source: FSC (2009).

demonstrates that southern members (55 percent) outnumbered northern members (45 percent) by the end of 2009.

As specified in the FSC bylaws (FSC 2002a), the General Assembly convenes at regular intervals, not to exceed three years. The chambers each elect three representatives for a three-year term to the board of directors, which is accountable to the membership. The General Assembly delegates most operational decisions and activities to the board and the FSC international secretariat, which was relocated from Oaxaca, Mexico, to Bonn, Germany in 2002. Although ultimate authority rests with the membership, the secretariat carries out the mandates of the membership, the strategic planning efforts of the board and the day-to-day operations of the program. The secretariat is led by an executive director, who is appointed by and is responsible to the board (FSC 2002a).

FSC has developed a global standard for its definition of 'well-managed forests', comprising ten principles and 56 criteria that cover key issues like land tenure and use rights to the land; indigenous peoples' rights; community relations and workers' rights; use of forest products and services; maintenance of biodiversity and high conservation-value forests; forestry planning, monitoring and assessment; and planning and management of plantations (FSC 2002b; see Box 3.1).[2] These general principles and criteria must be tailored to meet conditions in different countries through a process in which ecological, economic and social stakeholders have, in principle, equal decision-making powers. FSC has delegated authority to elaborate on its general principles and criteria to national affiliates. Consequently, national affiliates have considerable influence over the development of appropriate forest management rules in a national or regional context. Nationally or regionally developed standards are approved by FSC's board of directors if they conform to the scheme's

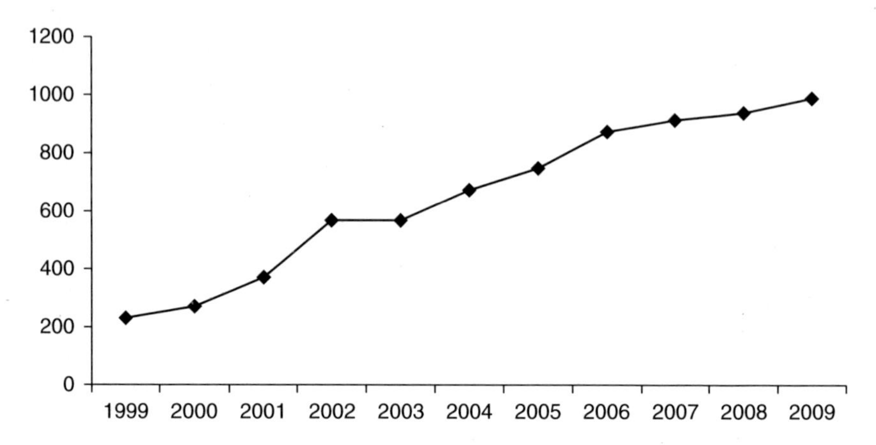

Sources: Auld (2009); data for 2008 and 2009 from FSC (2009) and www.fsc.org.

Figure 3.1 Number of FSC forest management certificates, 1999–2009

general principles and criteria and decision-making rules. In regions where there are no endorsed national standards, FSC-accredited certifiers may assess operations against standards that apply the general FSC principles and criteria appropriately for specific local conditions (FSC 2004). The number of endorsed national standards has not kept pace with the spread of forest management certificates, largely because developing national standards is a time-consuming and arduous process (Auld and Gulbrandsen 2010). As of December 2009, there were 57 national initiatives around the world, 30 standards endorsed by FSC's board of directors and almost 1000 active forest management certificates. Figure 3.1 shows the number of FSC forest management certificates from 1999 to 2009.

Because FSC arose through a lack of substantive results from intergovernmental collaboration on forest policies, its principles and criteria are not based on any regional set of intergovernmental criteria and indicators for sustainable forest management.[3] The lack of a linkage to intergovernmental criteria and indicators is due partly to timing. Work on the FSC principles and criteria began in 1991 (Synnott 2005), whereas intergovernmental criteria and indicator processes began after the 1992 Rio Earth Summit. The first intergovernmental process released a set of criteria and indicators in 1993, by which time FSC had moved through many versions of its principles and criteria and was close to completing them; they were finally ratified by the middle of 1994.

Another essential task of the program is the accreditation of third-party certifiers, which initially was handled by FSC's Accreditation Business

BOX 3.1: THE FSC PRINCIPLES

1. Compliance with laws and FSC Principles Forest management shall respect all applicable laws of the country in which they occur, and international treaties and agreements to which the country is a signatory, and comply with all FSC Principles and Criteria.

2. Tenure and use rights and responsibilities Long-term tenure and use rights to the land and forest resources shall be clearly defined, documented and legally established.

3. Indigenous peoples' rights The legal and customary rights of indigenous peoples to own, use and manage their lands, territories and resources shall be recognized and respected.

4. Community relations and worker's rights Forest management operations shall maintain or enhance the long-term social and economic wellbeing of forest workers and local communities.

5. Benefits from the forests Forest management operations shall encourage the efficient use of the forest's multiple products and services to ensure economic viability and a wide range of environmental and social benefits.

6. Environmental impact Forest management shall conserve biological diversity and its associated values, water resources, soils, and unique and fragile ecosystems and landscapes, and, by so doing, maintain the ecological functions and the integrity of the forests.

7. Management plan A management plan – appropriate to the scale and intensity of the operations – shall be written, implemented, and kept up to date. The long-term objectives of management, and the means of achieving them, shall be clearly stated.

8. Monitoring and assessment Monitoring shall be conducted – appropriate to the scale and intensity of forest management – to assess the conditions of the forests, yields of forest products, chain of custody, management activities and their social and environmental impacts.

9. Maintenance of high conservation value forests Management activities in high conservation value forests shall maintain or enhance the attributes which define such forests. Decisions regarding high conservation value forests shall always be considered in the context of a precautionary approach.

10. Plantations Plantations shall be planned and managed in accordance with Principles and Criteria 1 to 9, and Principle 10 and its Criteria. While plantations can provide an array of social and economic benefits, and can contribute to satisfying the world's needs for forest products, they should complement the management of, reduce pressures on, and promote the restoration and conservation of natural forests.

Source: FSC (2002b).

Unit. However, FSC was criticized for housing standard setting and accreditation in the same organization. The critics alleged that this arrangement could lead to a conflict of interest between the standard setters and those accrediting certifiers to assess compliance with the standards. In order to separate its accreditation function from its role as standard setter, FSC established Accreditation Services International (ASI) as a separate, independent organization that began operations in 2006. Still, the basic rules for accreditation of certifiers have been unchanged since 2004.

The largest certifiers include Société Générale de Surveillance (SGS), SmartWood, Scientific Certification Systems and the Soil Association. As noted, certifiers assess an operation against nationally or regionally developed FSC standards, but they may also assess operations against a 'generic' standard based on FSC's principles and criteria in countries where no such standards exist (FSC 2004). A forest company or landowner seeking certification must be approved in a major assessment conducted by a certifier. On passing this hurdle, the company receives an FSC certificate valid for five years and may sell the wood as certified, but forest operations are also monitored in annual audits. After five years of certified operations, the company must pass a comprehensive inspection, which is more wide-ranging than the annual audits. If forest companies or forest owners fail to correct serious certification standard shortfalls, they risk losing their certification. Certified producers can sell wood as FSC certified to processors, but using the FSC label on forest products sold to consumers also requires chain-of-custody certification, which involves tracking the origin of forest products throughout the supply chain. The chain-of-custody certificate allows forest products to carry the label if a certain percentage of the wood, chip or fiber contained in those products originates from FSC-certified forests.

The Emergence of Producer-dominated Schemes

FSC was quickly challenged by a number of programs initiated by forest industry and landowner associations. These programs initially operated under the strongly held belief that those who must actually implement standards for sustainable forest management – forest companies and forest owners – ought to develop the standards (Cashore et al. 2004). Several European forest associations responded to FSC by establishing landowner-dominated programs at the national level. In 1998 they joined to create the Pan-European Forest Certification Scheme (PEFC), to facilitate the mutual recognition of national programs and provide them with a common eco-label. The scheme was officially launched in 1999 by forest landowner associations in six European countries: Finland, Norway, Sweden, Austria, Germany and France. Unlike FSC, the PEFC regulations state that forest landowners are the appropriate initiators of the standard-setting projects: 'The process of development of certification criteria shall be initiated by national forest owners' organizations or national forestry sector organizations having the support of the major forest owners' organizations in that country' (PEFC 2006). Although 'all relevant parties will be invited to participate in this process' (PEFC 2006), it is largely up to the forest owners' organizations to define 'relevant parties' in standard-setting activities. The PEFC Council, composed of national governing bodies primarily representing forest owners' associations and other forestry organizations, approves national certification schemes if they are developed in conformity with the rules of the umbrella scheme. Unlike the case of FSC, there is no global set of principles and criteria stipulating requirements for forest operations. PEFC does have a set of basic requirements for national-level certification programs, first established in 2002 and revised many times since, but many of the specific requirements for forest operations are left to the discretion of national member programs. Thus, the program is of a different nature than FSC: whereas FSC is a global forest certification scheme with one set of principles and criteria, PEFC is a mutual recognition framework that endorses national certification schemes on the basis of certain requirements.

A similar development occurred in North America with the establishment of producer-dominated schemes in Canada and the USA. In 1993–4, the American Forest and Paper Association (AF&PA), the national association for the US forest industry, created the Sustainable Forestry Initiative (SFI) program as a more industry-friendly alternative to FSC (Cashore et. al 2004). The industry association launched SFI in 1994 and made adherence to its principles a condition of membership in the association, which represents 75 percent of US paper, wood and forest products. Originally an industry code of conduct, since its inception SFI has developed into a full-fledged certification program. Another scheme – the American Tree Farm System – was formed to provide a certification option tailored to nonindustrial landowners in the USA. This scheme certifies small landowners, whereas SFI is directed at larger operations. In Canada, a producer-backed initiative was announced during FSC's founding assembly in Toronto (1993), clearly indicating a mistrust of FSC (Elliott 1999).

The PEFC-endorsed European schemes use the Pan-European criteria and indicators for sustainable forest management as the framework for standard setting at the national level. Likewise, the US and Canadian certification programs claim to be fully consistent with the Montreal-process criteria and indicators. This claim may leave the impression that governments have established normative benchmarks for those certification schemes, despite the fact that criteria and indicators cannot be used to make claims about the stringency of a forest management standard (Rametsteiner and Simula 2003; Humphreys 2006). As noted, intergovernmental criteria and indicators, being primarily tools for information sharing about the state of forests, do not include targets or performance requirements for sustainable forest management.

In addition to being based upon governmental or intergovernmental criteria and indicators, the producer-backed schemes have developed linkages to government-sanctioned standard bodies and accreditation systems. These linkages are perhaps most visible in Canada, where the Canadian Pulp and Paper Association (CPPA), now the Forest Products Association of Canada, along with other industry associations, approached the Canadian Standards Association (CSA) about developing a forest certification program. CSA subsequently formed a multi-stakeholder technical committee to develop a forest certification standard. After drafting the standard, based on criteria and indicators developed by the Canadian Council of Forest Ministers, CSA adopted it in 1996 as the national standard for forest certification. With accreditation, the Canadian scheme also relied upon existing bodies endorsed by the government. Certifiers who audit CSA-approved forest companies must have qualifications approved by the Standards Council of Canada. The US-based programs, SFI and American Tree Farm System, also require certifiers to be accredited by existing government-sanctioned bodies – either the US Registrar Accreditation Board or the Canadian Environmental Auditing Association (AF&PA 2004).

The producer-backed schemes generally built certification procedures and program requirements from government-sanctioned criteria, indicators and guidelines in an effort to enhance their legitimacy and, arguably, to compensate for the lack of support from environmental NGOs (Cashore et al. 2004; Bernstein and Cashore 2007). Their standards were discretionary and flexible, focusing primarily on forest management, and, to a lesser extent, on environmental performance and social issues. Knowing that FSC had the higher standards, supporters of producer-backed schemes were quick to draw attention to the limited connections between FSC and intergovernmental criteria and indicators. They hoped that the producer-backed programs would compare favorably to FSC on this measure of legitimacy. Major differences between FSC and the producer-backed schemes are summarized in Table 3.2.

How can we briefly explain why industry and landowner associations developed FSC-competitor programs? Three such reasons can be identified in the literature on forest certification: the strong position of environmental and social stakeholders in FSC, the need for affordability and flexibility, and the need for an economic focus (Cashore et al. 2004; Gulbrandsen 2004).

First, perceptions that environmental and social interests dominate the decision-making process of FSC caused forest companies and forest owners to question its credibility and independence. In the absence of state legitimization, there is arguably an 'accountability deficit' in governance dominated by NGOs, because these organizations 'are not accountable for their actions in the sense required by even a minimalist theory of democratic governance' (Rosenau 2000, p. 192). FSC has taken this challenge seriously by strongly promoting the norms of accountability and transparency and by balancing the decision-making powers of economic, social and ecological interests. Nonetheless, a recurring theme among critics of FSC is usurpation, in that environmental NGOs are seen as self-appointed judges in a field in which they have inadequate understanding, limited experience and no legitimate right to regulate in the first place. This perception has moved forest owners and forest industries to establish producer-dominated schemes.

Second, the relatively high costs of FSC certification provided forest companies and forest owners with an incentive to develop more affordable and flexible standards. There are transaction costs related to pre-assessment studies (to determine the feasibility of certification) and the certification process itself, as well as costs involved in making on-the-ground changes

Table 3.2 Comparison between FSC and producer-backed programs

	FSC	Producer-backed programs
Initiation	WWF, other environmental NGOs and some socially concerned retailers	Forest landowner or industry associations
Rulemaking authority	Tripartite arrangement of economic, social and environmental stakeholders	Primarily economic stakeholders, although other stakeholders may participate
Stringency of standards	Relatively strict and non-discretionary standard	Discretionary and flexible standards
Scope of standards and auditing	Broad; forest management, environmental performance and social issues included	Narrower; forest management focus
Accreditation of certifiers	Independent organization created by FSC	National accreditation bodies
Linkages to intergovernmental criteria and indicator processes	None	Based upon regionally developed criteria and indicators
Geographic focus	Global program with national affiliates	National-level programs; endorsed by a mutual recognition framework

Source: Adapted from Cashore et al. (2004).

to become eligible for certification. In particular, the potentially high costs involved in complying with demanding certification requirements propelled companies and landowners to look for alternatives. In addition, the lack of a group certification option when FSC was launched (in which a single certificate covers a number of forest owners) motivated forest owners to develop more flexible certification options.[4]

Third, and closely related to the previous two issues, the stringency of FSC's environmental and social criteria motivated forest industries and forest owners to establish schemes that pay less attention to environmental and social criteria for sustainable forestry and more attention to economic criteria. The formation of producer-dominated schemes may be seen as an attempt to co-opt the discourse on forest certification and attract forest owners away from FSC's more stringent standards.

In summary, forest companies and forest owners in many countries considered FSC to be costly, intrusive and lacking in legitimacy. The result is that FSC and the producer-backed programs started a race to create support for their programs among wood retailers, wood processors and forest managers. FSC supporters claimed that it was the only credible certification scheme in existence, whereas the sponsors of producer-backed schemes argued that these programs were credible alternatives to FSC.

Mutual Recognition Discussions

The sponsors of producer-backed schemes initiated several processes in an effort to establish these programs as legitimate alternatives to FSC (see Cashore et al. 2004). This section will only review one of these processes. In 2001, the International Forest Industry Roundtable (IFIR), an informal network of forest companies and industry associations focusing on sustainable forestry, proposed a 'mutual recognition framework' for certification schemes; all schemes that passed an agreed threshold would be considered 'equivalent' (IFIR 2001). The proposal was justified on the grounds that it would resolve the problems emanating from the spread of competing certification schemes. Because of timber deliveries to the same manufacturers and paper mills from forest owners certified under different schemes, only a small fraction of the output could carry a label when it reached the market. For example, Swedish pulp and paper mills could not combine timber (chips and fiber) from FSC- and PEFC-certified forest and label the output with the FSC logo (see Chapter 5). If these schemes recognized each other as having equivalent standards, paper mills and manufacturers could combine timber flows from FSC- and PEFC-certified forests and sell forest products under the FSC label, or choose which label to use depending on market demand. No representatives of the producer-backed schemes had participated in the IFIR discussions, but they enthusiastically endorsed the idea of mutual recognition. FSC opposed the proposal, knowing that mutual recognition would erode its position as the only certification scheme supported by a wide range of NGOs and retailers. Its opposition to mutual recognition was supported by WWF, Greenpeace, Friends of the Earth and other NGOs, all of which claimed, with justification, that the proposal was an attempt by the forest industry to undermine FSC (Humphreys 2006, p. 132). The initiative failed when it became clear that FSC was unwilling to discuss mutual recognition of certification programs.

The IFIR proposal was later transformed into a proposal from the World Business Council on Sustainable Development (WBCSD) to reconcile FSC and the producer-backed schemes (Bernstein and Cashore 2004, p. 38). Referred to as the 'legitimacy threshold model', the proposal identified a number of factors as sources of legitimacy for certification programs (WBCSD 2003). Under this model, there would be not just one threshold above which all schemes are considered equivalent, but several 'legitimacy thresholds', corresponding to the needs of different user groups, which would allow for broader differentiation (Humphreys 2006, pp. 135–7). The legitimacy threshold model has been promoted since 2002 through the Forest Dialogue, a voluntary global partnership (Humphreys 2006, p. 135), but little has been achieved toward reconciling the various views of a credible forest certification scheme.

Following the failure of mutual recognition efforts at the international level, the producer-backed schemes have focused on mutual recognition among national-level schemes. As mentioned, PEFC was initially a mutual recognition framework for national-level schemes developed by European landowner associations. In 2003, PEFC restructured itself to become a global program, changing its official name to the Programme for the Endorsement of Forest Certification schemes, while retaining PEFC as its acronym.[5] The program has since endorsed the CSA program in Canada and the SFI program in the USA, as well as national-level certification programs in Brazil and Chile and, most recently, in Malaysia. Several other programs have associated membership but are not yet fully endorsed by PEFC. With this international expansion, the PEFC umbrella scheme became firmly established as a global competitor to FSC.

CERTIFICATION STANDARDS AND AUDITING PROCEDURES

Understanding the evolution of certification programs requires an examination of their changing standards and auditing procedures. This section reviews work that compares certification standards and auditing procedures across certification programs, demonstrating that the standards of producer-backed programs have become more stringent over time as a result of comparisons with the higher standards of FSC.

The Stringency of Certification Standards

Through the adoption of prescriptive environmental and social performance standards, certification could, *inter alia*, promote protection of old-growth forests, rare and threatened species and their habitats; restrict clearcuts and the use of chemicals in forestry; secure the rights of workers and indigenous peoples; enhance the wellbeing of local communities; and promote the sharing of benefits arising from the use of forests. Compared to FSC, the producer-backed schemes placed greater weight on standards of procedure, organizational and management measures, and flexibility in applying sustainable forest management standards (Rametsteiner and Simula 2003; Cashore et al. 2004). Environmental NGOs have repeatedly criticized forest landowner- and industry-led schemes for setting ecological and social standards that are too discretionary and lenient (for example Ozinga 2001, 2004; Vallejo and Hauselmann 2001). Not all schemes required on-the-ground field inspections of forestry operations and environmental performance, focusing instead on the management system of forest companies. This was initially the preferred approach of many producer-dominated schemes, but they have gradually introduced performance-based elements that require on-the-ground inspection (Cashore et al. 2004; Overdevest 2005).

There is significant variation in the stringency of standards both within and across programs. As of mid-2009, the PEFC Council has endorsed 25 nationally developed schemes (out of a total of 35 member schemes) that vary considerably in environmental and social rigor. Whereas some programs, such as those of Swedish forest owner associations, have environmental standards that are almost as stringent as those of FSC in the same regions, other European schemes have been criticized for their lax and discretionary environmental standards. The PEFC-endorsed SFI program in the USA and CSA program in Canada have also been criticized for prioritizing forest productivity and maximizing yield at the expense of environmental considerations (see for example Ozinga 2001 and 2004; Rametsteiner and Simula 2003). Partly as a result of the multilayered governance approach to standards development, regionally developed FSC standards also vary considerably in their environmental and social stringency. McDermott et al. (2008) found that differences among FSC's regional standards in the USA and Canada reflected variation in the relative prescriptiveness of underlying government regulations. To some extent, therefore, certification standards mirror regulatory and political conditions in the regions where they are developed. Some of the variation in the stringency of standards is due to differences in biophysical and socioeconomic conditions across regions, but there are also inconsistencies that can be explained only as variation in the ecological and social ambition of certification programs.

Because standard setting is a dynamic and iterative process, forest certification programs have changed considerably since their inception. The general pattern is that the standards of producer-backed programs have become more stringent over time (Cashore et al. 2004). Notwithstanding these changes, some basic differences between producer-backed programs and FSC are likely to persist. Detailed comparisons indicate that, in most countries and regions, FSC's ecological and social standards tend to be more stringent than are the standards endorsed by PEFC (Ramesteiner and Simula 2003; McDermott and Cashore 2008; McDermott et al. 2008; Auld et al. 2008; see also Chapter 5). Another difference that is likely to persist is the variation in comprehensiveness of the standards. Many producer-backed schemes have avoided dealing with social challenges in forestry, addressing only environmental and economic challenges. In this respect, their approach differs from that of FSC, which also focuses on social issues, including the rights of indigenous peoples and forest workers, the needs of local communities and equitable use and sharing of benefits derived from the forests.

The Certification and Auditing Processes

With regular on-the-ground forest auditing, certification programs could promote compliance with certification standards and continuous environmental and social performance improvements in forest management. Whereas the standard-setting body develops rules and accreditation requirements, independent certifiers conduct the actual certification of applicants.

As noted in Chapter 2, a salient difference among certification schemes is whether they are management system based (focusing on process) or performance based (focusing on outcome). Most certification schemes in forestry are based upon some combination of system-based and performance-based standards, but they place different weights on these differing types of standards. The principles and criteria of FSC embody relatively stringent performance-based standards, which require on-the-ground field inspections. A FSC certification process typically includes preliminary assessment; on-the-ground field inspection by a team of professional foresters, biologists and other experts; consultation with local communities; preparation of a preliminary assessment report by the certification body and peer review of the report; discussion with the applicant; a final certification determination and issuance of a certificate; and annual follow-up audits (Meidinger 2006, pp. 70–71). Rejection of applicants seeking certification is rare; the certifier usually issues a list of corrective action requirements (CARs) that a producer must address before being certified (precondition) or soon after being certified (condition). FSC requires certifiers to disclose certification and audit reports to the public.

As of 2005, the scheme has also allowed for the participation of observers in the certification process and has developed rules for their participation: how observers are selected and accepted, their role in the certification process, requirements related to confidentiality and observer conduct and costs (Meidinger 2006, pp. 81–2). Hence, the possibility of observer participation in the certification process is a part of the transparency requirements of the program (Auld and Gulbrandsen, 2010).

Although producer-backed programs tend to have less inclusive and transparent auditing procedures than FSC does, their auditing procedures and practices have improved over time. The US-based SFI program, for example, was initially a code of conduct with no requirements for third-party auditing. The program has been revised several times, however, adding a number of auditable criteria and indicators (Cashore et al. 2004; Overdevest 2005). Similarly, since establishing its basic requirements in 2002, PEFC has adopted new conventions to enhance transparency, stakeholder participation and third-party monitoring (PEFC 2006, 2007, 2009). Nonetheless, there has been considerable reluctance within the PEFC-endorsed schemes to include environmental and social stakeholders in the certification process or to consider complaints from them (Nussbaum and Simula 2005). These schemes also provide less public information on inspections than FSC does (Meidinger 2006, pp. 81–2). Schemes that do not publish audit reports make it difficult for interested stakeholders to verify claims about the effects of the auditing process. Moreover, it can be assumed that the less stringent and precise the standards, the greater the scope for differing interpretations by auditors (Gulbrandsen 2004, p. 90). We could therefore expect that the impact of FSC audits would be greater than the impact of audits in producer-dominated schemes with more discretionary standards. The environmental and social impact of on-the-ground auditing is discussed in Chapter 4.

Explaining the Increasing Stringency of Standards and Auditing Procedures

Many observers expected that the emergence of producer-backed programs with discretionary and weak requirements would erode certification standards, especially as industry and landowner associations were successful at enrolling their constituencies in their programs. Contrary to this expectation, we see that competition with FSC for credibility and support has caused poorer-performing producer-backed programs to increase the stringency of their standards. In a study of the emergence of forest certification schemes in Canada (British Columbia), the USA, the UK, Sweden and Germany, Cashore et al. (2004) found that many producer-dominated schemes have responded to criticism from FSC supporters by 'changing upward'. Since the inception of the Swedish PEFC, for example, it has adjusted its standard upward, almost to the level of stringency of the Swedish FSC standard, through mutual recognition efforts and competition with FSC (see Chapter 5).

Much like Fung et al. (2001) argue for the importance of public comparisons in their work on ratcheting labor standards, work on forest certification has identified the strategic use of public comparisons between certification programs to exert upward pressure on the standards and governance procedures of producer-backed initiatives. Overdevest (2005) found that strategic comparisons with the higher standards of FSC, often orchestrated by environmental NGOs, played a key role in the increasing stringency of SFI standards in the USA. Reports and matrices that compare schemes over a range of criteria (for example Ozinga 2001 and 2004; Vallejo and Hauselmann 2001) were used to pressure SFI and other producer-backed programs to increase the stringency of their standards. According to one NGO-sponsored comparison, for example, SFI could not be accepted as a credible certification system, owing to its weak standards, lack of transparency and unbalanced governance processes (Ozinga 2001). In response to the NGO-sponsored reports, some buyers hired consultants to undertake 'official' criteria-by-criteria comparisons of certification schemes. In the USA, for example, Home Depot commissioned the Meridian Institute to do a comparison of SFI and FSC (Overdevest 2005). The report (Meridian Institute 2001) found substantial differences in their procedural, substantive and governance standards, demonstrating that the producer-backed programs were the poorer performers. In Europe, the Confederation of European Paper Industries (CEPI) produced a similar report, comparing certification schemes on a range of criteria (CEPI 2001). The effect of these comparisons was a ratcheting up of producer-backed programs, narrowing the gap between their approach and that of FSC (Cashore et al. 2004; Overdevest 2005). As discussed in Chapter 5, similar reports were produced in Sweden and Norway, resulting in an upward change in PEFC-endorsed national programs in both countries.

Public procurement requirements in the UK, the Netherlands, Germany, France, Denmark and other European countries have also made a significant contribution to the increasing stringency of producer-dominated programs. A growing number of governments have adopted procurement programs, stipulating the purchase of forest products from legal and sustainable sources. These governments have considered certification under credible schemes to be evidence of the legality and sustainability of forest products. Public procurement requirements in the UK have been particularly influential in forcing producer-backed certification schemes to change their standards. Because research on forest certification has rarely focused on public procurement policies, the remainder of this section will provide an examination of the impact of UK procurement policies on certification programs.

The UK government procurement policy was announced in 2000, when the Minister of the Environment announced that all government departments and their agencies would 'actively seek' to buy timber and timber products from legal and sustainable sources.[6] An inter-departmental working group was established to advise on and monitor implementation of the policy, but progress on the delivery of the government's promise was slow. Following a series of embarrassing incidents in which Greenpeace revealed that the Home Office and other departments were using illegally logged timber, the government came under pressure to tighten its policies and provide additional implementation support and guidance (ENDS 2003, p. 39).

In 2003, the UK Department for Environment, Food and Rural Affairs (DEFRA) appointed the consulting company, ProForest, to run a Central Point of Expertise on Timber (CPET) to advise the UK government on timber procurement. The consultants first developed assessment criteria for forest certification programs in order to evaluate (1) an assurance of legality and (2) an assurance of sustainability. They then assessed five certification programs against the criteria: FSC, PEFC, the North American scheme (SFI), the Canadian scheme (CSA) and the Malaysian Timber Certification Scheme (MTCS). The consultants' report, published in November 2004, concluded that timber certified by FSC and the Canadian scheme should be accepted as assurance of legal and sustainable timber (CPET 2004). Certificates from PEFC, the US-based SFI scheme and the Malaysian scheme should be accepted as assurance of legal timber, but not of sustainable forest management. PEFC and the Malaysian scheme did not meet the public procurement requirements because of unbalanced governance, inadequate public consultation during the certification process and lack of public disclosure of auditing outcomes. The North American scheme, SFI, was not approved because the chain-of-custody certificate did not specify the amount of uncertified material used in the product (CPET 2004).

The schemes that did not pass the test were allowed six months, beginning November 2004, to improve their standards and rules before DEFRA implemented preferential treatment (ENDS 2004a, p. 32). In response, PEFC addressed each of the CPET issues at its 2004 General Assembly and at a second, especially scheduled, meeting in April 2005. Subsequently, PEFC moved to adopt new conventions on balanced governance, public consultation and transparency.

In August 2005, DEFRA announced that PEFC and SFI had improved their standards sufficiently to meet the public procurement requirements; but they were on probation until the end of 2005, when they were to be reassessed by CPET. Greenpeace, Friends of the Earth and other NGOs immediately attacked DEFRA's decision, saying in a joint statement that the UK government had approved two schemes that 'allow large-scale, unsustainable logging in ancient forest areas, the destruction of endangered species and the abuse of indigenous people's rights' (ENDS 2005, pp. 22–3). Notwithstanding such NGO protests, CPET's reassessment confirmed that PEFC and SFI – like FSC – could be used as evidence of legal and sustainable timber, whereas the Malaysian scheme did not pass the sustainability test (CPET 2006).

In summary, just as Vogel (1995) has argued that trade partners sometimes voluntarily adopt higher environmental standards, producer-backed certification programs have harmonized up to access buyers in green markets. Public procurement policies and side-by-side comparisons of standards have forced the public exposure of information about certification rules and practices, thereby pressuring the more poorly performing programs to improve their standards. Public comparisons with FSC's higher standards effectively forced the producer-backed schemes to increase the stringency of their standards and to enhance participation from outside stakeholders in standard-setting and certification processes.

MULTI-STAKEHOLDER VERSUS PRODUCER-DOMINATED GOVERNANCE MODELS

Through its tripartite decision-making structure and open membership model, FSC created a capacity and commitment to be accountable to a wide range of environmental, social and economic stakeholders. The success of FSC in attracting widespread support among market players, NGOs and governments seems to have legitimized and authorized an organizational model based on the participation of a wide range of parties, shared decision-making authority and empowerment of those affected by the actions of power wielders (Gulbrandsen 2008). This recipe in turn incorporates and is reinforced by global norms about transparency, corporate social responsibility, stakeholder democracy and deliberations between business and civil society. Over time, we see evidence of some degree of convergence among certification programs and institutional *isomorphism* – the tendency toward organizational homogeneity (cf. DiMaggio and Powell 1983).

Most of the PEFC-endorsed programs have evolved from initiatives owned and operated by industry or landowner associations to become fully independent, third-party certification programs. In this process, they have become much more isomorphic with FSC. Industry and landowner associations continue, however, to dominate standard-setting processes. Although PEFC-endorsed programs have opened up to some participation from environmental and social stakeholders, economic stakeholders remain in control of rulemaking and governance. In the PEFC Council, voting rights are based on the size of the forest owners' land, and environmental and social stakeholders have no formal voting rights. The result of the governance reforms in producer-backed programs is arguably a governance model that seems to be open to participation from environmental and social stakeholders, whereas, in fact, it restricts their influence in actual decision-making processes. Consultation with environmental and social groups could be a way of justifying actions without needing to be answerable to any stakeholders other than industry peers. It would do little to improve an organization's actual responsiveness to external audiences. It is not surprising, therefore, that despite efforts to strengthen the independence of producer-backed programs, most environmental NGOs have little confidence in them. Instead of endorsing the steps taken by those programs to strengthen their independence, WWF and other environmental groups responded by intensifying campaigns to promote FSC in the marketplace. Most environmental NGOs considered the efforts of producer-backed schemes to increase the participation of environmental and social stakeholders in their governing bodies as strategic moves to make them appear inclusive and independent without fundamentally altering responsiveness to environmental criticism or making it easier for environmental groups to influence rulemaking and governance.[7] They claimed that those programs reserve for themselves the right to decide who is accountable to whom and for what. Accountability understood in this way could become a meaningless ritual of justifying conduct by answering only those questions that the answerable party has decided upon (Pellizzoni 2004).

On the other hand, many forest owners did not agree that they ought to be accountable to environmental and social stakeholders. They were skeptical of FSC precisely because of its inclusiveness in rulemaking and certification processes, contending that environmental and social groups should not wield significant rulemaking influence, given that these groups bear no responsibility for implementing the rules or any of the costs of complying with them. Their opposition to FSC was based not solely upon the stringency of the FSC standards, but also upon the groups that were supporting the program (Cashore et al. 2004, p. 234). Against this background, we can identify two distinct governance arrangements in non-state certification programs. In one, forest owners and forest companies are primarily accountable to their peers; in another forest owners and forest companies are accountable to multiple environmental, social and economic stakeholders. These two models have persisted, despite steps taken within producer-backed schemes to strengthen the independence of their governing bodies.

CONCLUSIONS

The origins of certification go back to the concerns of environmental NGOs about the tropical deforestation and biodiversity loss of the 1980s and their pressure on large retailers to avoid the sourcing of tropical wood. Lack of effective government regulations was a significant factor in making environmental NGOs turn to the industry itself. By creating FSC in partnership with other stakeholders, the NGOs circumvented arduous intergovernmental forest policy negotiations and established a non-state certification standard. FSC was based upon relatively stringent standards and required broad stakeholder participation in standard development, decision-making transparency, accreditation of third-party auditors and public disclosure of certification reports.

In response to FSC, industry and landowner associations moved quickly to establish FSC-competitor programs with more discretionary and flexible standards. Because of their success in attracting forest companies and landowners, many observers expected to see a race to the bottom, whereby certification would be eroded as a credible instrument of environmental governance. Yet, the evidence shows that producer-backed programs have been forced to adjust their standards upwards as a result of public comparisons with the higher standards of FSC. As a result, a new form of governance has emerged, different from traditional state regulations and various forms of industry self-regulations and voluntary measures. Forest certification schemes are well positioned to achieve a high standard of accountability relative to other governance experiments such as international codes of conduct or industry self-regulatory schemes, which do not involve prescriptive standards or third-party auditing of compliance.

The process of increasing the stringency and autonomy of producer-backed schemes will likely be difficult to reverse, and could, in the long run, enhance the organizational capacity for responsiveness to external constituencies. But the producer-backed certification schemes have thus far avoided decision-making rules and structures that could significantly reduce the influence of forest owners and forest industries in standard development and operation. The steps taken by those schemes to improve auditing and open up to some participation from outside stakeholders can be explained as strategic adaptations to market demands. The adoption of new conventions on verification, auditing and stakeholder consultation could, in fact, serve to justify the business-as-usual situation. In the absence of real stakeholder participation and influence, procedural arrangements such as consultation with stakeholders and independent auditing of practices may do little to enhance responsiveness to constituents such as environmental organizations and local communities. Organizations may strategically adopt popular, taken-for-granted organizational recipes in order to divert criticism of their activities, rather than acting to improve their conduct. It is necessary, therefore, to look beyond the formal procedures of non-state certification schemes to explore the impacts of certification. This critical issue is examined in the next chapter.

NOTES

1. Unlike trade, procurement is a member state competency in the EU (Humphreys 2006, p. 158).
2. FSC originally had nine principles; the tenth, on plantations, was added in 1996. For more on the principles and criteria, see the FSC website: www.fsc.org.
3. Although FSC's principles and criteria are not based on intergovernmental criteria and indicators, the ITTO *Guidelines for the Sustainable Management of Natural Tropical Forests* (ITTO 1992) served as a foundation for early drafts of FSC's principles and criteria (Synnott 2005).
4. In 1998, FSC adopted a group certification option.
5. See the PEFC website: www.pefc.org.
6. www.press.dtlr.gov.uk/pns/DisplayPN.cgi?pn_id=2000_0516.
7. Interviews with representatives of WWF and Friends of the Earth in 2001, 2003 and 2005.

4. The adoption and impact of forest certification

This chapter reviews what we know about the adoption, direct effects and broader consequences of forest certification. The dramatic growth in land areas certified during the past 15 years attests to the market success of forest certification. This chapter demonstrates, however, that it is necessary to consider patterns of adoption when assessing the effectiveness of such voluntary tools as certification. There is a risk that only producers who face relatively low adoption costs opt in, whereas producers who face substantial adoption costs systematically opt out. This selection problem continues to raise questions about effectiveness.

The chapter begins with a review of the formation of buyer networks to promote forest certification among retailers and manufacturers of wood products. The second section examines the market penetration of certification programs, detailing the impact of demand-side support for certification. The third section examines patterns of adoption and some of the unintended consequences of certification. The fourth section turns to NGO criticism of and support for forest certification. The fifth section reviews the extant research addressing on-the-ground impact of the auditing process, and discusses the potential of certification to reduce pressure on high conservation value forests and pressure for deforestation. The conclusion reflects on the effectiveness of forest certification as an instrument of environmental and social governance.

THE FORMATION OF BUYER NETWORKS

Beginning in 1991, Friends of the Earth, various rainforest action groups and other environmental activists began targeting do-it-yourself retailers in the UK. Meanwhile, WWF spearheaded the establishment of professional buyer networks in order to create demand for certified products. In 1991, WWF-UK established the world's first buyer group – the WWF 95 Group – comprising retailers committed to increasing the volume of wood products from well-managed sources. The name was derived from the members' pledge to support WWF's ambitious goal, set in 1989, for the world's tropical timber trade to be based on sustainable timber sources by the end of 1995. At the time, one of the difficult issues for environmental groups was retailers' misuse of claims about wood products sourced from well-managed forests. Because FSC was not yet operational, the original requirements for membership committed retailers to an immediate phase-out of all wood labels and certificates claiming sustainability until 'a credible independent certification and labeling system' was established (Murphy and Bendell 1997, p. 109).

Over the next few years, the group attracted many retailers that pledged adherence to its objectives. Some members, the home improvement retailer B&Q in particular, showed an immediate interest in supporting FSC in order to verify that its wood products were sourced from well-managed forests. As explained by Alan Knight, B&Q's lead purchaser: 'We weren't losing customers (. . .) But we knew that if our name, B&Q, was associated with destruction of tropical forests or even temperate forests, that our brand name would be damaged' (quoted in Counsell and Loraas 2002, p. 12). Others joined the 95 Group after direct targeting from more radical environmental groups such as Greenpeace, Earth First and Friends of the Earth, which organized a series of protests at retail outlets and wholesalers' timber distribution centers (Cashore et al. 2004, p. 141). The campaigns proved to be highly successful, attracting considerable public attention and pressuring retailers to sign up with the WWF buyer group.

Initially focusing only on do-it-yourself retailers and the hardwood timber trade, the 95 Group expanded its reach to cover other sections of the forest industry, including pulp and paper, furniture and construction. In January 1995, the group adopted new requirements for membership, supporting FSC as the only credible certification and labeling system. Members were required to phase out the purchase of wood products that did not come from FSC-certified forests, 'demonstrably increasing' the proportion of certified wood by the end of 1995. A major boost came immediately after the agreement on the new membership criteria, when two of the UK's major supermarket chains – Sainsbury's and Tesco – joined the group, increasing its turnover of wood products to over £2 billion (Murphy and Bendell 1997, p. 115). Yet, many members fell short of the target to increase the proportion of FSC-certified wood because of limited supply. At the end of 1995, WWF and its partners agreed to continue working together to increase the volumes of FSC-certified wood, changing the name of the group to the 95 Plus Group (Murphy and Bendell 1997).

Similar buyer groups were established in several other European countries. In the Netherlands, for example, the Hart voor Hout (Heart for Wood) campaign was launched in 1992 by Milieudefensie (Friends of the

Earth Netherlands) and Novib, a development NGO. Similar to the UK experience, direct action against do-it-yourself retailers resulted in commitments to stop selling tropical timber products, with the exception of timber products certified under a credible scheme. Although the Dutch section of WWF was not a formal partner, Hart voor Hout campaigners maintained close relations with forest program officers from the organization (Murphy and Bendell 1997, p. 130). By 1997, buyer groups had been established in six European countries – the UK, the Netherlands, Germany, Switzerland, Belgium and Austria – and in North America (UNECE/FAO 1999).

In several countries, support from large retailers was forthcoming only after intensive NGO targeting, threats of boycotts and shaming campaigns. In the USA, the Rainforest Action Network initiated a series of protests against Home Depot, accusing the company of selling wood products sourced from endangered old-growth forests. Between 1997 and 1999, the activists targeted Home Depot outlets with public protests, letter-writing campaigns and publicity stunts: 'rappelling from roofs, chaining themselves to piles of old-growth wood, and arranging logging slash to write the Home Depot logo onto clear-cut hillsides' (Klooster 2005, p. 408). In March 1999, the Rainforest Action Network and other groups coordinated simultaneous protests in 150 Home Depot stores across the USA. In August 1999, Home Depot's CEO, Arthur Blank, announced that the company would stop selling wood products from 'environmentally sensitive areas' and that it would 'give preference to wood that is certified as coming from forest managed in a responsible way' (quoted in Conroy 2007, p. 71). Under the new policy, Home Depot agreed to purchase wood certified by FSC 'or equivalent' schemes, clearly indicating that it preferred FSC (Cashore et al. 2004, p. 112). The Rainforest Action Network responded by running a full-page ad in the *New York Times*, thanking Home Depot for its decision and encouraging its competitors to follow suit. The second largest home improvement retailer in the USA, Lowe's, moved quickly, announcing its own procurement policy with support for FSC. Within months, most of the leading home improvement retail chains in the USA announced similar commitments (Conroy 2007, p. 75). Once aligned, the large retailers became allies of NGOs, calling for producers to certify so they could meet the demand for certified wood.

MARKET PENETRATION

With substantial market penetration, certification programs have been able to promote international trade in wood products sourced from well-managed forests. WWF has currently established buyer groups in over 30 countries, coordinated through its Global Forest and Trade Network (GFTN). Originally a demand-side network, as of mid-2009 GFTN included producers and buyers that controlled more than 20 million hectares of certified forests, employed nearly 3 million people globally, and traded 16 percent of all forest products traded internationally (WWF 2009). The network's policy is that FSC certification is the only evidence of credibly certified forest products.

In 2009, industrial roundwood production from certified sources – both the FSC and PEFC – totaled 411 million m^3, accounting for about 26 percent of the world's annual production (UNECE/FAO 2009). Yet, only a small fraction of this wood carries a logo when it reaches the market (Auld et al. 2008, p. 193), largely because of low consumer awareness and lack of a price premium on labeled products. In general, there is greater demand from retailers and consumers for the FSC label than for the PEFC label.

Disaggregating by markets reveals important patterns: FSC has the upper hand in the timber market, whereas PEFC is strongest in the pulp and paper market (UNECE/FAO 2009). Many industrial pulp and paper companies demand certified material, but no particular label. The acceptance of FSC-competitor programs in the pulp and paper market is evidence that the effectiveness of a specific scheme is dependent upon widespread producer adoption of the scheme. As noted, without large volumes of certified wood in the market, it is almost impossible for large buyers to enforce a specific procurement policy, whether self-imposed or collectively agreed upon within buyer networks. FSC has grown rapidly, however, reducing the problem with inadequate supplies of certified wood. Furthermore, because of the increasing stringency of standards in producer-dominated programs, they have become acceptable to many retailers and governments that initially signaled a preference for FSC-certified wood.

The structure of the global commodity chains for forest products facilitated the rapid spread and expansion of certification initiatives: a relatively small group of giant retailers buys a significant proportion of the forest products traded internationally (Klooster 2005, p. 408). The demand for certified wood from such large retailers as Home Depot, Lowe's, IKEA and B&Q effectively convinced many producers to certify. Rather than representing a costly tool, certification provided the retailers with a useful instrument for exerting 'control at a distance' over suppliers (Ponte and Gibbon 2005, p. 22; Klooster 2005; Raynolds et al. 2007). In this sense, the emergence of certification programs proved to be a win-win situation for many retailers; they avoided NGO targeting, while shifting the costs of quality control and monitoring onto their suppliers. Their primary challenge, particularly in the early years of certification, was that FSC was too poorly established to supply them with sufficient volumes of certified wood. The losers, in economic terms, were the producers, who had to pay for the certification costs without gaining a price premium on certified wood. Unlike Fair Trade certification of coffee and other products, forest certification schemes neither guarantee producers a minimum price for wood nor require wood processors or retailers to invest in social development. Wood prices are determined by market mechanisms, and large retailers typically require not only certified wood but also high volumes and low prices (Klooster 2005). Niche and high-end products are more amenable to price differentiation, however – particularly native tropical woods certified by FSC (Auld et al. 2008, p. 194).

As discussed in Chapter 3, governments have also been critical on the demand side. Although FSC explicitly forbids the participation of governments in standard development, the initiative was supported by several governments that saw eco-labeling as a way of circumventing trade rules that hindered them from imposing tropical timber import restrictions to control illegal and irresponsible logging. The UK, the Netherlands, Denmark, Germany and France pioneered the development of public procurement of wood products from legal and sustainable sources. These and other European governments, as well as Japan, have identified certification by a credible scheme as a way of verifying that public procurement requirements for legal and sustainable timber are met (Gulbrandsen and Humphreys 2006), attesting to the importance of interplay between private and public governance. Public procurement policies have not only facilitated market acceptance of certified forest products, but have also enhanced the rulemaking legitimacy of several forest certification schemes. By approving specific certification schemes, governments are signaling that those schemes are legitimate and credible governance systems that private procurers and other buyers can trust.

Market penetration is limited largely to Europe and North America. Unlike PEFC, FSC has recently added a number of logo users in Japan and China, but Chinese logo certificates exist primarily to service FSC demand in Europe and the USA (UNECE/FAO 2007). To date, demand for certification in developing countries has been modest. Forest holdings in tropical countries have had little trouble selling uncertified and even illegally sourced timber on the world market. In addition to this obstacle to the proliferation of certification in tropical countries is the fact that only a relatively small proportion of the industrial roundwood harvested in tropical forests enters international trade (Rametsteiner and Simula 2003, p. 96). We see evidence, however, of increasing interest in certification in a number of tropical countries. Certification is not only accepted in the marketplace, but is also a requisite of many European and US buyers of tropical timber and wood products. Forest cooperatives in developing countries are also motivated by the hope that certification may enable them to compete in international markets in a manner similar to producers of Fair Trade products (Klooster 2006). This means that demand-side support is likely to continue to stimulate the proliferation of forest certification in the tropical region. Like private sector demand, public procurement policies and the EU's Forest Law Enforcement, Governance and Trade (FLEGT) scheme (see Chapter 3) have increased interest in certification in tropical countries. We can therefore expect to see a further expansion of certification programs in the tropical region over the next decade.

PATTERNS OF ADOPTION

With participation from a critical mass of producers, certification programs could have a significant impact on forestry practices across countries and regions. The certified forest area worldwide has increased steeply since the inception of FSC in 1993. By May 2009, FSC- and PEFC-certified lands totaled 325 million hectares or approximately 8.3 percent of the world's forest cover (UNECE/FAO 2009). Canada, the Russian Federation, Sweden and the USA account for nearly 62 percent of all FSC-certified lands. Although FSC has certified forestland in a number of developing countries, those forestlands provide small contributions to the total area certified by the program. Table 4.1 provides an overview of FSC certified forest areas around the world as of December 2009, showing that 46 percent of total FSC certified areas was in Europe and 37 percent in North America. The center of PEFC's activity, even more than FSC's, is in Europe and North America. Canada alone accounts for 40 percent of the area certified with PEFC endorsement; Nordic and other European countries account for another 28 percent; whereas the two developing countries with PEFC-endorsed schemes – Brazil and Chile – add only about 1 percent each to the PEFC total (Auld et al. 2008, p. 193). Figure 4.1 shows the forest area certified by major certification programs from 1999 to 2009.

The certified forestland in developing countries represents about 10 to 20 percent of the world's total certified area (UNECE/FAO 2009). Most of the tropical countries are lacking any type of certification scheme. Explanations have indicated the costs of certification; lack of knowledge about certification programs; uncertain or disputed land tenure; and incompatibility of laws, traditional rights and certification standards

Table 4.1 FSC certified forest area and certificates around the world, as of December 2009

	Certified area (million hectares)	Percent of total certified areas	Number of certificates
Europe	54.38	46	443
North America	43.36	37	197
Latin America and Caribbean	9.90	8	206
Africa	5.53	5	43
Asia	3.25	3	81
Oceania	1.51	1	27
Total	117.93	100	997

Source: FSC (2009).

(OECD 2003; Fisher et al. 2005; Cashore et al. 2007a). As noted in Chapter 3, the prescriptiveness of certification standards mirrors the stringency of underlying governmental forest regulations (McDermott et al. 2008). The broader point, however, is that certification is more likely to be adopted in countries with relatively stringent government regulations than in countries with weak government regulations, thus widening the gap that separates the environmental management capacities of developed and developing countries.

Another unintended consequence of certification is the favoring of large operations and forest companies at the expense of nonindustrial owners practicing small-scale forestry. Because of economies of scale, it is easier for the larger forest companies to participate in certification programs and comply with management rules. According to Klooster, FSC certification has evolved from a niche-market tool to become 'a document intensive, buyer-driven preoccupation for delivering large quantities of certified wood products to the market, with a focus on big forest producers and large wood consumers' (Klooster 2005, p. 412). Since 1998, however, FSC has offered a group certification option to reduce costs for small forest owners. Group certification allows a number of small owners to join together and share certification costs. In principle, the number of group members is unlimited, but FSC requires groups to be managed effectively and according to its rules, which implies that they cannot be too large.

The problem of high transaction costs for small owners has also spurred efforts to form specialized programs to reduce entry barriers for smallholders. This problem partly explains the formation of PEFC by European

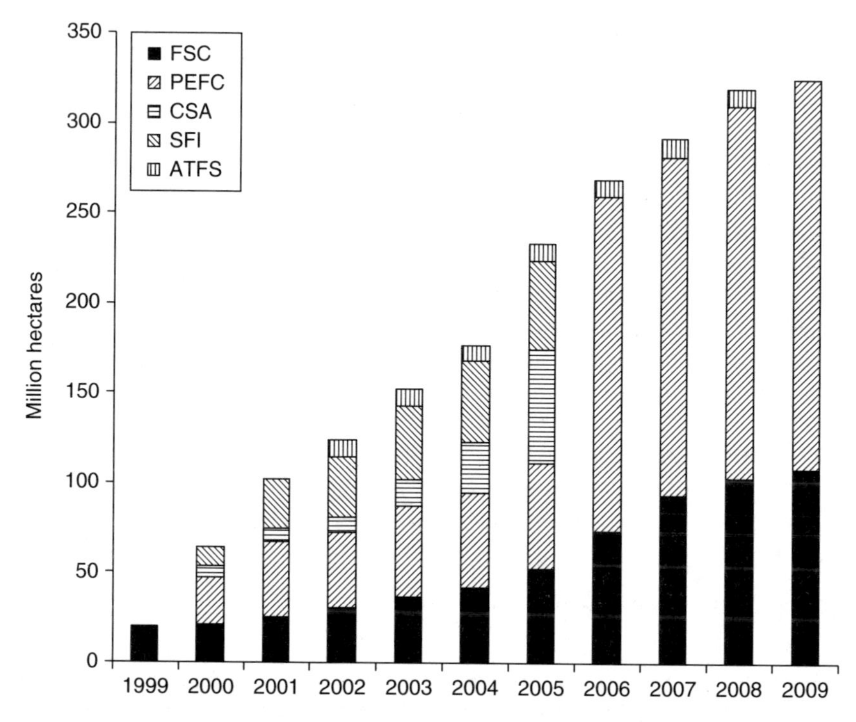

Note: Certification schemes indicated here are Forest Stewardship Council (FSC), Programme for the Endorsement of Forest Certification Schemes (PEFC), Canadian Standards Association's forest certification scheme (CSA, endorsed by PEFC in 2005), Sustainable Forestry Initiative (SFI, endorsed by PEFC in 2005), and American Tree Farm System (ATFS, endorsed by PEFC in 2008).

Sources: Individual certification schemes, UNECE/FAO (2009) and author's compilation.

Figure 4.1 Forest area certified by major certification schemes, 1999–2009

forest owners and the American Tree Farm System by forest owners in the USA (Cashore et al. 2004), but neither of these programs has assisted small-scale forest owners in developing countries. FSC, on the other hand, began work in 2001 on a social strategy to improve smallholder access to the program and to address the needs of forest communities, indigenous peoples and forest workers (Klooster 2005, p. 412). This strategy focused on strengthening social standards in forestry and reducing entry barriers for small and low-intensity forest users. In 2002, FSC started work on specific guidelines for certification of small and low-intensity managed forests (SLIMF). These guidelines were developed to lower the bar for certifying smallholders (less than 1000 hectares of forests), low-intensity management and harvesting operations, and producers of non-timber forest products. Although these corrective measures indicate that FSC, after years of rapid expansion, has taken steps to improve access to its certification program, the overall picture is that large forest companies and operations rather than small-scale forest owners are still the dominant economic stakeholders in the program.

An examination of participation in competing schemes shows clearly that FSC has not become the one and only global standard-setting body for market-driven certification that environmental organizations had hoped for. In 2002, PEFC overtook FSC in total certified area (Rametsteiner and Simula 2003). By the end of 2009, PEFC-endorsed schemes had certified about twice as much forestland as that certified by FSC. Despite the emergence of producer-backed competitors, however, FSC has continued to grow at an impressive rate, and consumer recognition of its label is increasing.

NGO CRITICISM AND SUPPORT

Although FSC is supported by a wide range of environmental and social NGOs, such support is qualified and conditional. Many NGOs opposed the controversial 1996 decision to include a principle on plantations, but criticism of FSC has since focused largely on issues of implementation (Humphreys 2006, p. 129). The most wide-ranging and thorough critique of FSC came from the Rainforest Foundation, a nonprofit organization dedicated to protecting tropical forests (not to be confused with the Rainforest Alliance), with the 2002 publication of the voluminous report, *Trading in Credibility* (Counsell and Loraas 2002). Based on case studies of implementation in Brazil, Thailand, Malaysia, Indonesia, Ireland and Canada, the Rainforest Foundation alleged that FSC certification was hampered by inadequate audits, lack of effective control mechanisms and other serious flaws. The report found that there had been difficulties in implementing FSC Principle 2 (tenure and use right and responsibilities) and Principle 3 (indigenous peoples' rights). It also found that FSC's 'fast growth' strategy had resulted in certification of noncompliant forest operations and that it undermined multi-stakeholder processes in the cases investigated. Because certifiers compete for clients (certification applicants) in the market, there were 'vested corporate interests' between certifiers and their clients in ensuring successful certification outcomes (Counsell and Loraas 2002, p. 5). To avoid the alleged collusion and manipulation between certifiers and forest managers, the report recommended that

FSC should cease to accredit independent certifiers and instead employ assessors to conduct the certification of forests, with certification fees paid directly to FSC.

Regarding governance, the report claimed that FSC functioned poorly as a democratic membership-based organization; key stakeholders were 'effectively excluded' from FSC processes at international, national and local levels; local communities and indigenous peoples remained 'marginalized' in decision-making processes (Counsell and Loraas 2002, p. 7). Although FSC, backed by WWF, responded that many of the allegations were untrue, inaccurate or out of date, they agreed that there were areas for improvement and stated that they were taking steps to address those concerns (FSC 2003). The Rainforest Foundation still does not recommend tropical timber approved by any certification scheme, alleging that the public cannot be assured that a certified forest product comes from a well-managed forest. Interestingly, material for the Rainforest Foundation report would have been impossible to compile without FSC's transparent governance processes and public disclosure of policy documents, certification decisions and assessment reports.

FSC members occasionally criticize the program, but they rarely go public with their critiques. In October 2006, a number of NGOs wrote a letter to FSC's executive director and international board and called for governance reforms, expressing deep concern about the loss of credibility that FSC was allegedly suffering. Among the signatories were both members and nonmembers of FSC, including WWF, Greenpeace, Birdlife, Friends of the Earth UK, the Sierra Club, Environmental Defense and the Rainforest Foundation. A copy of the letter was not made publicly available, but its content is summarized at FSC-Watch, a webpage dedicated to publishing critical reports and examples of FSC certifications.[1] As with the 2002 Rainforest Foundation Report, the signatories of the letter alleged that certifiers were eroding FSC's credibility because of the direct financial link between them and the companies they certify.[2] They claimed that certifiers had too much discretion in interpreting FSC's principles and criteria and that the FSC secretariat had failed to control and sanction certifiers who breached the rules. Accordingly, they recommended that FSC find an alternative business model that would break the direct financial link between certifiers and their clients. This letter shows that even WWF and other ardent supporters of FSC sometimes join forces with hard-nosed 'outsiders' like the Rainforest Foundation to call for governance reforms. FSC has not ceased to use independent certification bodies, but, as noted in Chapter 3, it has established an independent accreditation organization (ASI) in order to separate its accreditation function from its role as standard setter.

Notwithstanding the occasional criticism from some groups, support for FSC remains high within most environmental and social NGOs. As discussed, most NGOs have focused on criticizing the producer-dominated programs, while backing FSC as the only credible certification program. According to the NGOs, the producer-backed programs are less democratic and transparent than FSC is, leaving them with little scope for scrutinizing forestry practices and providing forest owners with great leeway in applying certification standards. A recent Greenpeace report praises FSC, concluding that the program 'is in effect an elaborate conflict resolution mechanism for reconciling many differing views and values in relation to forests and some plantations' (Rosoman et al. 2008). Three facets of the program appear to be important in fostering this kind of support from a wide range of environmental and social NGOs.

First, FSC's three-chamber structure means that the environmental and social chamber combined controls two-thirds of the votes in the General Assembly, whereas the economic chamber controls only one-third of the votes. In this respect, the program clearly empowers social and environmental interests in decision-making processes over and above most other transnational bodies.

Second, FSC's choice to devolve authority to make a global standard locally appropriate to national affiliates was important to stakeholders' perception of the organization. It meant that environmental and social stakeholders at the national and local level were systematically included in standard-setting activities. It also meant that some of the controversy over appropriate forest management standards was effectively transferred to lower governance levels in the system, leaving the General Assembly free of arduous discussions that could have had a negative impact on the program. Indeed, the development of national and regional standards has taken upwards of eight years, thus being among the most time-consuming activities within the program (Cashore et al. 2004). Devolving authority to national affiliates thus served the dual purpose of democratizing the program by granting decision-making rights to stakeholders at the national level, while reducing conflict at the level of international governance.

Third, the decision to address not only environmental issues, but also social issues in FSC's principles and criteria appears important in fostering NGO support for the program. From the beginning, FSC recognized the need to establish long-term tenure and user rights of forestland (Principle 2), to respect the legal and customary rights of indigenous peoples (Principle 3), to maintain or enhance the wellbeing of forest workers and local communities (Principle 4) and to encourage the use of forests to ensure economic viability and a wide range of environmental and social benefits (Principle 5). This focus on social issues meant that those concerned over the consequences of destructive industrial logging for indigenous peoples, local communities and traditional forest workers had reason to support FSC. As noted, FSC has been criticized for prioritizing fast growth at the expense of achieving its social objectives, but it did respond to such criticism by developing a social strategy that better served the needs of small landowners, forest workers, indigenous peoples and local communities. In this respect, FSC differs from producer-backed programs, which tend to focus more on the economic needs of forest companies and forest landowners.

THE IMPACT OF ON-THE-GROUND AUDITING

Although a detailed review of the on-the-ground impacts of certification is beyond the scope of this chapter, some general findings based on existing literature can be identified.[3] Extant research has focused mainly on the effects of FSC certification, largely as a consequence of the lack of information disclosure on audits from producer-backed programs. Studies of the distribution of corrective action requirements (CARs) issued by certification bodies indicate that auditing does have an impact on forest management practices. As mentioned in Chapter 3, CARs are conditions that must be met before or shortly after certification. Early reviews of CARs issued by FSC-accredited certifiers indicated that most changes pertained to forest planning, documentation and monitoring, as opposed to on-the-ground practices (Rametsteiner 1999; Bass et al. 2001). In another study, Rametsteiner and Simula (2003. p. 95) noted that 'it can safely be said that forest certification has brought along improvements in internal auditing and monitoring in forest organizations'. Concerning the on-the-ground impacts of forest certification, they were careful to avoid premature conclusions, but indicated that auditing is likely to make forest managers more sensitive to issues like natural regeneration, thinning operations, reduced-impact harvesting, forest road construction, use of chemicals and relations with society (Rametsteiner and Simula, 2003). They also noted that high variability in the quality of auditing could be observed among different certification schemes and different auditors of the same scheme.

A more recent study of 80 audits in the United States by SmartWood, an FSC-accredited certifier, found that 94 percent required improved management plans, 79 percent required enhanced monitoring, 71 percent dealt with inventory issues and 69 percent involved forest mapping activities (Newsom et al. 2006). Although these figures show that 'system elements' predominate as requirements specified by CARs, the study also found significant attention to ecological issues. Regarding forest protection, 79 percent of operations were required to address management of sensitive sites and high conservation value forests,[4] and 63 percent had to address other ecological issues such as protection of threatened and endangered species. Similar studies in Sweden and Norway have reported similar results (see Chapter 5).

With social issues, work on the impact of FSC certification shows that certifiers address the conditions of forest workers and local communities more regularly in developing countries than they do in developed countries. In 2005, the Rainforest Alliance issued a report analyzing a sample of 129 SmartWood audits in 21 countries (stratified by regions: South America, Central America and Mexico, Asia, New Zealand and Australia, United States and Canada and Europe). The analysis showed that certifiers more often address social issues like training, worker safety, worker wages and living conditions, and conflict resolution with local communities and other stakeholders in developing countries (Newsom and Hewitt 2005, p. 22). Similarly, studies of forest certification in Brazil (Espach 2006), Bolivia (Nebel et al. 2005) and Mexico (Klooster 2006) confirm that certifiers more regularly address compliance with laws and regulations of workers' rights in developing countries (Auld et al. 2008, p. 198). In Brazil, for example, operations seeking FSC certification in order to gain their certificate had to comply with government laws that were largely unenforced (Espach 2006).

Whether or not certification has reduced pressure on high conservation value forests and the pressure for deforestation is an issue that has undergone debate. The work on CARs previously referred to shows that forest operations had to change practices around sensitive areas and high conservation value forests. When we turn to the issue of environmental impact, however, the link to certification programs becomes more tenuous. As Auld et al. (2008, p. 198) explain, we must consider what protection on individually certified tracks means for forest protection at the landscape level. They note that forest protection on individually certified tracks, which leads to reduced harvesting, can mean higher pressure for timber extraction on noncertified lands. The patchwork of adoption raises questions about the capacity of certification to address protection at the landscape level. Indeed, researchers agree that landscape-level planning is necessary to address concerns such as wildlife conservation and the management of large predators (Bennett 2001; Putz and Romero 2001) and the appropriate placement of plantation forests versus areas for ecological protection (Sedjo and Botkin 1997). These issues of scale represent a significant hurdle in using certification as a tool for addressing environmental problems that are rarely contained within a single forest (Auld et al. 2008, p. 199). With regard to reducing pressure for deforestation, there is broad recognition that potential price premiums on certified timber or greater market access following certification provide an inadequate counterbalance to stronger economic incentives for land-use conversion (Gullison 2003; Auld et al. 2008).

About half of the total area certified by FSC in developing countries comprises plantations – typically uniform monocultures of fast-growing softwood with little genetic variability. This is a paradox, given that when certification was first proposed to the ITTO in the late 1980s, the idea was to certify tropical, mega-diversity forests, but certainly not plantations. Wood products originating from plantations are marked with the same eco-label as naturally grown forests; customers have no means of distinguishing wood products from naturally grown forests. Although plantations may take the pressure off commercial utilization of naturally grown forests, the problem is that in order to facilitate faster growth, natural forests are often replaced with plantations. FSC does not permit certification of plantations that are converted from naturally grown forests, but in a long-term perspective most plantation forests were once natural forests. With so few certified naturally grown forests in the tropics, we should not expect forest certification to halt the rate of deforestation, tropical forest degradation and loss of biodiversity. Moreover, as Dauvergne (2001) found in a study of corporate forestry practices in the Asia-Pacific area, the informal and political nature of state–business relations in the region represents a considerable barrier to change in logging practices on the ground. Thus, he concluded, although transformation of the formal principles and rules of forest management has increased the pressure on loggers to modify practices, it is not likely to save the remaining old-growth commercial forests (see also Dauvergne 2005). Notwithstanding substantial differences between Southeast Asia, Latin America and Africa in the causes leading to deforestation, researchers agree that wide-ranging policy interventions and socioeconomic reforms in all regions are necessary to protect the tropical forests (see for example Lambin and Geist 2003). The scale of deforestation in the tropics means that certification on individual tracts can hardly make a significant impact (Gullison 2003).

The upshot is that forest management practices are changing as a result of FSC audits. The findings show that although implementing environmental standards in forestry practices is a serious challenge, certifiers do address breaches of the standards and FSC-certified forest owners need to demonstrate progress in dealing with the problems in order to keep their certification. However, certification has been unable thus far to ameliorate environmental deterioration in forestry significantly. The ability of certification to ensure the conservation and sustainable use of biological diversity is unclear. There is also broad recognition that certification cannot be the tool for addressing forest protection at the landscape level or reducing pressure for deforestation (Gullison 2003; Auld et al. 2008). On balance, although we do see evidence that FSC certification has changed forest management practices, certification is still not an effective environmental institution in the sense of resolving some of the most pressing environmental problems in forestry.

CONCLUSIONS

Market and producer adoption of forest certification has been driven by NGO targeting of companies along wood product commodity chains. Following environmental group campaigns and WWF's establishment of buyer groups to create market demand for FSC, large retailers in Europe and North America demanded supplier documentation, proving that wood products came from sustainably managed sources. The proliferation of forest certification has not been driven by consumer demand or the hopes of a price premium on certified wood. Nonetheless, environmental NGOs would certainly have had less success in their efforts to create markets for certification without the threat of consumer boycotts. The fact that retailers were aware of the ability of consumers to express political and ethical preferences through boycotting specific brands or products – the danger of which environmentalists brought home to them – mattered more than did actual buying behavior.

The emergence of producer-backed schemes was probably the most significant unintended consequence of the creation of FSC (Auld et al. 2008). Widespread adoption of such schemes has marginalized FSC in several countries, reducing its chances of setting a truly global standard for well-managed forests. Patterns of adoption thus confirm, in some measure, the claim of an inherent conflict between two necessary conditions for effectiveness in voluntary, market-based instruments: the need for strong environmental standards and the need for widespread producer participation. Unless markets are prepared to pay a significant premium for strong labels, producers will, not surprisingly, tend to prefer labels under schemes with weaker and more flexible standards. On the other hand, in the case of forest certification, FSC has continued to grow, and producer-backed programs have increased the stringency of their standards following competition from FSC and pressures from environmental NGOs and market players. This development has extended the impact of certification more broadly.

It remains to be seen if forest certification can solve the dire environmental and social problems in forestry, or if it amounts to little more than a successful marketing tool. Although certification standards are diverse and audits are of varying quality, the evolution of substantial and wide-ranging certification schemes shows that many forest owners and forest companies are prepared to go beyond legal requirements. In this chapter, audited operations were shown to have been required to change on-the-ground practices to participate in schemes and to address the protection of high conservation value forests. Yet, extant research demonstrates skepticism about the ameliorative potential of certification for landscape-level protection and deforestation. Landscape-level planning and policy interventions seem necessary to address environmental problems that are rarely contained within an individually certified tract. Moreover, despite tropical forest degradation being a major reason for creating certification initiatives, adoption of certification programs in developing countries is modest compared to adoption in developed countries. Although interest in forest certification is increasing in developing countries, the market benefits accruing from certification have evidently been insufficient to convince large numbers of forest holdings to undertake the necessary steps to become certified. In addition to the financial cost of certification, there is little knowledge of certification programs and control of forestland, and a large proportion of the forestland that is certified in the tropics comprises plantations.

NOTES

1. www.fsc-watch.org/.
2. www.fsc-watch.org/archives/2006/11/06/Green_groups_call_for_urgent_reform_of_the_FSC_certifiers_eroding_credibility (accessed 29 June 2009).
3. This literature review draws on collaborative work with Graeme Auld and Connie McDermott (Auld et al. 2008).
4. High conservation value forests (HCVF) is an FSC term to describe forest with environmental and social values of significant importance.

5. Forest certification in Sweden and Norway

Understanding the emergence and impact of forest certification requires a study of the way certification processes play out in specific political and socioeconomic contexts. This chapter examines the emergence and effects of forest certification in Sweden and Norway. These two forest-rich Scandinavian countries have been selected not only for their many similarities, including forest biodiversity and ecology, dependence upon paper product exports, administrative traditions, and relationships among business, non-governmental actors and the state, but, crucially, for differences in their forest industry structure. The story told in this chapter demonstrates that the structure of the forest sector influenced industry and landowner responses to certification pressures and their adoption choices, as well as the unfolding of the standard-setting process. The processes investigated also reveal that certification choices at critical decision points created path dependencies that shaped and constrained the unfolding of certification processes.

This chapter begins with an examination of the formation of forest certification programs in Sweden and Norway. The second section compares the stringency of certification standards, demonstrating that the poorer performers have increased the stringency of their standards over time. The third section examines patterns of adoption and a discussion of mutual recognition of certification schemes in Sweden, where two competing certification programs exist. The fourth section reviews evidence that sheds light on the crucial question of what is known about the on-the-ground environmental impact of forest certification. The fifth section turns to the broader consequences of certification, examining whether or not collaboration among environmental, economic and social stakeholders in standard-setting processes has resulted in shifting alliances and new cleavages. In closing, the sixth section links the specific discussion of certification in Sweden and Norway to broader concerns about the effectiveness of this instrument in transnational environmental governance.

THE FORMATION OF FOREST CERTIFICATION SCHEMES

The Swedish forest sector is larger and more export oriented than the Norwegian forest sector. Sweden has about 23 million hectares of productive forestland and a forested area of more than 28 million hectares, covering 66 percent of the whole land area. In Europe, only Russia has a larger forested area. Forest products exports comprise 12 percent of the Swedish export income, and the forest sector accounts for 12 percent of total employment in Swedish industry. Six industrial forest companies control 39 percent of the Swedish forestland. Sveaskog, a state-owned company formerly known as AssiDomän, is the largest in forest ownership (15 percent of the Swedish forestland). Some 51 percent of the forestland is owned by approximately 336 000 nonindustrial private owners, of which one-third is a member of four regional landowner associations (Norra Skog, Norrskog, Mellanskog and Södra). The rest of the forestland is owned by other owners, including municipalities and the Swedish Church (Swedish Forest Agency 2008). The ownership of forests in Sweden varies among regions. Nonindustrial private owners own most of the forestland in southern parts of the country, whereas the forest companies own a large share of the forestland in northern Sweden.

Norway has more than 7 million hectares of productive forestland and a forested area of about 12 million hectares, covering 38 percent of the whole land area. There is only one major pulp and paper company in Norway, and it owns little forestland. Approximately 80 percent of the forestland is owned by the 120 000 nonindustrial private owners, of which about one-third belong to eight regional landowner associations. The remaining 20 percent is divided among the state and municipalities (12 percent of the total), industry and companies (4 percent) and other private owners (4 percent). Like Sweden's forest industry, the Norwegian forest sector is dependent upon export markets. Forestry played a major role in Norway's economy a century ago; yet the current turnover in the forest industry accounts for merely 5 percent of the total turnover of Norwegian industry.[1]

The remainder of this section examines the formation of FSC in Sweden and the emergence of landowner-dominated programs, in both Sweden and Norway. As this section demonstrates, both countries were frontrunners in the adoption and implementation of FSC or landowner-dominated programs.

NGO-driven Process in Sweden

Environmental NGOs have worked long and hard to protect old-growth forests and forest biodiversity in Sweden. For many years, their strategy was to target governments and influence public forest policy through such measures as increasing the number and size of government-protected forests and campaigning for strict provisions to protect biodiversity in forest policy. In the last two decades, however, they have added a new tactic to their repertoire: targeting the forest products' supply chain. Many of the large environmental NGOs, most notably WWF, Greenpeace and Friends of the Earth, have national and local groups around the world, and are thus well placed to promote forest certification in the global marketplace. As discussed in Chapter 4, WWF-UK established a network of buyers (WWF 95-group) to support sustainable forest management among manufacturers and retailers in 1991. Buyers in the group had to commit themselves to phasing out timber from poorly managed forests. When FSC was legally registered in 1995, WWF required members to sign a commitment to give preference to FSC-certified wood. Some powerful members of the group, such as do-it-yourself retailer B&Q, went even further in their support for FSC, and publicly announced their intention to phase out all non-FSC-certified timber.

Similarly, Nordic environmental NGOs created the Taiga Rescue Network to promote protection of the northern boreal forests and to enable quick communication, allowing them to organize actions in several countries simultaneously. This network established its headquarters in northern Sweden in 1993, and is said to have successfully pooled the resources of a large variety of environmental groups with the common goal of protecting coniferous forests (Sæther 1998, pp. 189–90).

In Germany, Greenpeace targeted domestic publishing houses in 1993, accusing them of using printing paper originating from old-growth or ill-managed forests in the Nordic countries. Springer Verlag, the giant German publisher, responded by asking its suppliers to document whether or not old-growth forests were set aside and harvesting operations were sustainable. As a result of the NGO pressure, in 1995 the Association of German Paper Producers and the Association of German Magazine Publishers issued a position paper signaling a preference for products purchased under a credible global program that could be implemented rapidly (Cashore et al. 2004, pp. 170–71). Although they did not require suppliers to participate in FSC, there was little doubt that the big German publishers had FSC-style certification in mind, as they expressed in several meetings with their suppliers (Mäntyranta 2002).

The new strategy devised by the environmental NGOs was apparently much more effective than were old-style government lobbies. The NGOs had to balance their act, however; while exerting pressure on forest owners by upping conflict levels, they wanted forest owners to cooperate with and participate in FSC in the development of standards for well-managed forests. In 1994, WWF Sweden established an advisory 'reference group' of scientists and forestry stakeholders to work out a set of criteria for biodiversity conservation in Swedish forestry. A year later, the group issued a set of preliminary certification criteria. Although some forest companies wanted to proceed with the development of an FSC certification standard, many key players in the forestry trade were skeptical of what was perceived as the rigorous demands of the environmental organizations.[2] The forestry industry tried to establish a Nordic Forest Certification Project, but that initiative failed – largely because it was boycotted by the major environmental NGOs in Sweden, Finland and Norway.

Instead of supporting the Nordic initiative, WWF and the Swedish Society for Nature Conservation established a Swedish FSC working group in which a written declaration of support for FSC's general principles and criteria was a requirement of membership.[3] These organizations were joined by various stakeholders, including Greenpeace and Friends of the Earth (environmental interests); retailers and the Swedish Church in its capacity as a forest landowner (economic interests); and the Swedish Sami Association and labor unions (social interests). As a result of publicly announced preferences for FSC-certified wood products by powerful buyers in the UK and Germany, the large forest companies, AssiDomän (renamed Sveaskog in 2002) and Korsnäs, decided to undergo FSC pilot testing in 1995.[4] These companies declared that they would join the Swedish FSC working group even if the other major forest companies decided not to participate (Elliott 1999, p. 383). Under pressure from AssiDomän and Korsnäs, and, perhaps more importantly, customers in Germany and the UK, the other forest companies decided in 1996 to have the Swedish Forest Industries Federation represent all the Swedish forest companies in the FSC working group. According to the director of the Federation, both market pressure and a strong tradition of speaking with one voice influenced the decision of the other members of the industry association.[5] This decision helped tip the scale for the regional forest owner associations representing the nonindustrial private forest owners, and they agreed to join the working group collectively (Gulbrandsen 2005a, Cashore et al. 2004).

The NGOs largely set the working group's agenda (Elliott 1999, pp. 385–9). This situation was accepted by the forest companies, but led to resentment in the forest owner associations.[6] Sami demands concerning reindeer grazing, along with relatively high environmental standards and uncertainties concerning group certification options, divided the associations. Specifically, indigenous use of forests has divided opinion in Sweden following the expansion of forestry in the 1980s to areas traditionally used for reindeer herding in the northwestern part of the country (Hellström

2001, pp. 33–5). The Sami argued that the Swedish FSC standard must recognize their customary rights, whereas private forest owners in the north felt that modern reindeer herding, which involved moving reindeer by trucks, had a negative impact on their forestland, and could not fall under customary rights (Klingberg 2002, cited in Cashore et al. 2004, p. 204). The industrial forest companies did not oppose Sami demands, which were directed primarily at nonindustrial forest owners in northern Sweden. In May 1997, the forest owner associations agreed collectively to leave the process. Greenpeace also withdrew from the working group, because it was strongly opposed to some intensive harvesting methods that were acceptable to the other participants in the group. By the end of 1997, the remaining environmental NGOs, Sami representatives, forest companies and the other working group members reached agreement on a Swedish FSC standard that subsequently became the first nationally developed standard approved by FSC's international board (Elliott and Schlaepfer 2001, p. 645).

Immediately following the nonindustrial forest owners' withdrawal from the FSC working group, it was realized within the forest owner association, Södra – the only association with industrial facilities in Sweden – that an alternative to FSC had to be developed in order to prevent loss of market shares to FSC-certified companies. Led by Södra, the Swedish forest owner associations joined forces to develop a forest certification standard, defining the environmental performance level of their members' forestry practices. Using conformity with these standards as a requirement for certification, the associations offered group certification to their members (Gulbrandsen 2005a).

Landowner-driven Process in Norway

In 1994, the Norwegian Forest Owners' Federation, representing the regional forest owner associations, and the forest industry had informal discussions on establishing a project that could build international and domestic confidence in the declaration that the raw materials from the forest industry were based on sustainable forestry practices.[7] They officially launched the Living Forests project in 1995 to develop a national standard for sustainable forest management and build environmental skills among forest owners. Ecological, economic and social interests were represented in equal measure in the standard development group, but Sami representatives did not participate, because reindeer herding by the Sami in Norway occurs primarily in areas of little value to forestry. The main players in the working group were the Norwegian Forest Owners' Federation and the forest industry, WWF Norway and the Norwegian Society for Nature

Conservation, outdoor and recreational interests, labor interests and representatives from the Ministry of Agriculture and the Ministry of the Environment. The Ministry of Agriculture, which funded 25 percent of the project costs, required the Living Forest standard to conform to the criteria, indicators and operational guidelines of the Pan-European ministerial conferences on forests.[8] In 1997, a certification committee was established as part of the Living Forests project to consider different systems, including FSC, the International Organization for Standardization's (ISO) environmental management system (EMS) standard ISO 14001, and the EU Eco-management and Audit Scheme (EMAS).

After almost three years of intensive negotiations, all participants in the Living Forests project agreed in 1998 on a national standard comprising 23 requirements for sustainable forest management. The certification committee had not ranked any of the schemes to certify forest owner associations, preferring instead to show how the Living Forests standard could be used in combination with other systems. The standard could be applied almost without further elaboration in combination with the management systems-based ISO 14001 or EMAS schemes, whereas FSC certification would require further elaboration to adapt the standard to its international principles and criteria. WWF and the Norwegian Society for Nature Conservation proposed that a Norwegian FSC working group should be established to elaborate and expand upon the Living Forests standard.[9] This step would require a new process by which implicated economic, environmental and social interests – including Sami representatives – would participate on an equal footing, but the forest owner associations rejected the proposal, and opted for ISO 14001 systems-based certification. Hence, the forest owners decided on their own which type of certification scheme they would adopt. An undertaking certified according to ISO 14001 is required to have an environmental policy and goals in place, but can basically decide on the environmental performance level for itself. Furthermore, ISO 14001 is process-oriented, meaning that an organization can be certified before it fulfils certain criteria, as long as it can demonstrate improvements from one assessment to the next.

The forest owner associations developed a group certification system for their members in accordance with ISO 14001 and the Living Forests standard. In 1998, the first forest owner association in Norway offered a group certificate to all its members, and within three years, all the forest owner associations in Norway had followed its lead. Most of the timber traded in Norway is brokered through the associations that negotiate timber prices, buy timber from the forest owners and sell it to the industry. Forest owners supplying timber through the associational system are required to comply with the Living Forests standard. Those who refuse to participate collectively through their associations lose traditional trading channels for timber. Thus, the carrot of increased sales and low transaction costs encourages participation and the stick of reduced sales discourages defection (Gulbrandsen 2005a).

By certifying their management systems in accordance with nationally developed forest management requirements and a system-based standard such as ISO 14001, Norwegian and Swedish forest owner associations obtained a combination of performance and system-based forest certification schemes. Although certification to a system-based standard lent some credibility to the forest owners, it was obvious that certified timber and wood products needed an internationally recognized logo attesting to sustainable forestry. In the absence of a credible alternative to FSC, the forest owner associations in Norway and Sweden, in partnership with landowner associations in Finland, Austria, Germany and France, forged an alternative: the PEFC scheme, launched in 1999. With PEFC's endorsement of the Nordic 'family forestry' schemes in 2000, all certified associations in Norway were included in one international scheme, while two competing international schemes gained a foothold in Sweden. Table 5.1 summarizes the key differences between the Swedish FSC and the Norwegian PEFC programs.

THE STRINGENCY OF CERTIFICATION STANDARDS

The Norwegian Living Forests standard (Living Forests 1998) initially appeared to be less prescriptive and comprehensive than did the Swedish FSC standard (FSC Sweden 1998). Perhaps the most salient difference was that the Swedish FSC required that at least 5 percent of the productive forest be permanently set aside, whereas compliance with the Living Forests scheme resulted in conservation of approximately 1 percent of the forestland. The Swedish FSC requirements were stricter than the Living Forests requirements with regard to registration, handling and protection of natural forests and habitats with red-listed species. With the exception of forest road construction, the same could be said of requirements concerning ditching, preservation of dead wood and unproductive forestland, exotics, the use of herbicides and the rights of the Sami people. Table 5.2 compares the original Swedish FSC and Norwegian Living Forests standards on salient environmental and social issues.[10]

Like the Norwegian standard, the Swedish PEFC standard (PEFC Sweden 2000) used to be more discretionary and flexible than the Swedish FSC standard. In Sweden, as in other European countries and North

Table 5.1 A comparison of the Swedish FSC and the Norwegian PEFC programs

Salient dimensions	Swedish FSC	Norwegian PEFC
Initiation	WWF Sweden and Swedish Society for Nature Conservation	Norwegian Forest Owners' Federation and the forest indusry
Rulemaking authority	Tripartite arrangement of economic, social and environmental stakeholders	Tripartite arrangement of economic, social and environmental stakeholders, but forest owners decided on certification system
Type of certification	Individual (certification of each forest management unit)	Collective (group certification)
System operation	Based on Swedish FSC standard	Based on Living Forests standard and ISO 14001
Stringency of sustainable forest management standards	Relatively strict and wide-ranging environmental and social standards	More flexible and discretionary environmental and social standards
International orientation	Based upon FSC's international principles and criteria; scope for use of FSC logo	Pan-European criteria and indicators; endorsed by PEFC, scope for use of PEFC logo

America (see Chapter 3), strategic comparisons with the higher standards of FSC contributed to a ratcheting up of PEFC-endorsed standards. A European NGO, Fern, produced a series of reports called *Behind the Logo*, comparing the standards and procedures of FSC and national competitor schemes in Sweden, Finland, Germany, France, Canada and the USA. Based on a side-by-side comparison of standards and procedures, the Swedish *Behind the Logo* study (Lindahl 2001) concluded, not surprisingly, that FSC was more stringent than PEFC on a number of criteria. WWF Sweden followed Fern's example and produced a Swedish-language report called *Bakom kulisserna* (Behind the Scenes) (Dahl 2002) that corroborated those findings. As discussed later, an independent

Table 5.2 The stringency of the original Swedish and Norwegian certification standards

Standards	Swedish FSC (1998)	Norwegian Living Forests (1998)
Set-aside areas	At least 5 percent	Approximately 1 percent
Natural forests and key habitats	Natural forests and key habitats to be registered and preserved	Qualities of natural forests and key habitats to be sustained
Ditching	New ditching to be prohibited	New ditching permitted if no harm to biologically valuable mires and wetlands
Dead wood	Dead wood to be protected from forest measures; standing dead wood to be left when thinning and regeneration felling	Large windfalls left on the ground over five years to be left in forest
Forest road construction	No specific requirements	Restrictions on the construction of roads on biologically valuable forestland
Unproductive forestland	Land-use change not to be permitted	Afforestation to be permitted; drainage ditching on bogs and forest wetland to be avoided
Chemicals	Chemical pesticides and herbicides that are harmful to the environment and health not to be used	Herbicides not to be used in silviculture when clearly more efficient than mechanical methods
Genetically modified organisms	Prohibited	Prohibited
Exotics	Minimal use; permitted only in exceptional cases following consensus decision in FSC Board	Permitted where natural regeneration is too slow to yield economically sustainable harvest

Table 5.2 (continued)

Standards	Swedish FSC (1998)	Norwegian Living Forests (1998)
Outdoor life	Public access to forests must be maintained	Public access to forests must be maintained
Indigenous peoples	Sami people's rights recognized	No specific requirements beyond following government regulations

Sources: FSC Sweden (1998); Living Forests (1998).

criterion-by-criterion comparison, jointly commissioned by the supporters of FSC and PEFC, demonstrated that the Swedish FSC was the most stringent and prescriptive standard on a number of key issues (Aulén and Bleckert 2001). In response to these reports, the Swedish PEFC Board initiated a standard revision process. A revised PEFC standard was approved in December 2004 by the Swedish PEFC Board and endorsed by the international PEFC Council in February 2006. In the standard revision, PEFC Sweden adjusted its standard upward to match the requirement of FSC Sweden on such ecological issues as set-aside areas (at least 5 percent), retention trees and landscape planning (PEFC Sweden 2006). The program also adopted new rules to enhance transparency, labor rights and third-party monitoring (PEFC Sweden 2006).

Even in Norway, where FSC had not gained a foothold, strategic comparisons with the higher standards of FSC resulted in a ratcheting up of the PEFC-endorsed Living Forests standard. In the absence of a Norwegian FSC standard, the Swedish FSC standard provided Norwegian NGOs with a suitable benchmark for comparison with the Living Forests standard. WWF produced a Norwegian language report (Håpnes and Hvoslef 1999) that concluded that FSC was the most prescriptive and comprehensive standard (see also Table 5.2). In 2003, in response to NGO and market concerns that the Living Forests standard was too discretionary, the forest owners initiated a standard revisions process in collaboration with WWF and other stakeholders. The revised 2006 version of the Living Forests standard is more stringent and comprehensive than the original 1998 standard, narrowing the gap between its approach and that of FSC. Like the Swedish FSC, for example, the revised Living Forests standard requires that 'at least 5 percent of productive forest areas shall be managed as areas of ecological importance' (Living Forests 2006, p. 11).

The most significant difference between FSC and PEFC in Sweden and Norway now appears to be the participation in and influence of environmental and social stakeholders in governing the programs: environmental, social and economic interests have, in principle, equal decision-making influence in FSC, whereas forest owners still tend to dominate rulemaking processes in PEFC. Moreover, FSC seems to have more transparent governance, certification and auditing processes. These issues are discussed in greater detail in the section on mutual recognition discussions.

ADOPTION OF CERTIFICATION SCHEMES

This section turns to the issue of certification program adoption in Sweden and Norway, beginning with a common measure of direct effects: the area certified by schemes and the proportion of certified to uncertified land. From these descriptive statistics, the section examines the question of how we can explain patterns of adoption. The section ends with a review of mutual recognition efforts in Sweden.

Patterns of Adoption

All major forest companies (Sveaskog, Holmen, Stora Enso, SCA Skog, Bergvik Skog and Korsnäs), as well as other landowners such at the Swedish Church, municipalities and governmental authorities, are certified according to the Swedish FSC standard, comprising about half of the productive forestland (almost 11 million hectares out of approximately 23 million hectares of productive land). The Swedish forest owner associations offer group certification under PEFC to their members on a *voluntary* basis: forest owners who wish to participate sign a contract with their association. The result of voluntary group certification is clearly reflected in the statistics: only a relatively small proportion of the forestland owned by members of the associations is certified (varying from about 2 to 20 percent of the land owned by members of the four associations). By mid-2009, PEFC had issued 22 group certificates and three individual certificates in Sweden. In total, about 3 million hectares of forestland owned by nonindustrial forest owners is certified under PEFC in Sweden. As discussed later in this chapter, four out of the six FSC-certified forest companies have also become PEFC-certified in recent years (Holmen, Stora Enso, Bergvik Skog and Korsnäs), thus increasing the total area certified under PEFC in Sweden to 7.5 million hectares.

Unlike the Swedish forest owner associations, the Norwegian associations made group certification *mandatory* for their wood suppliers in order to maximize adoption of the system.[11] Because forest owners who sell timber through the associational system must comply with the certification standard, the forest owner associations claim that all forestland controlled by small forest owners in Norway – comprising about 80 percent of productive forestland – is certified. This system is accepted under PEFC and endorsed by its council. Most of the remaining 20 percent of the productive forestland has become certified under the same system and endorsed by PEFC.

According to official PEFC statistics, all the commercially productive forestland in Norway – comprising more than 7 million hectares of forests – is certified. As noted, the timber traded through the forest owner associations must be sourced from certified operations, but the claim that all productive forestland in Norway is certified can be questioned. Rather than certifying all forest management units, the Norwegian system is based upon the forest owner associations being certified in accordance with ISO 14001 and the Living Forests standards. Individual forest owners who supply wood to the associations must comply with the certification standards, but those who do not sell timber through the associational system are not necessarily required to adhere to the standards in logging operations.

Explaining Patterns of Adoption

The Swedish forest companies are all vertically integrated, with their own industrial facilities (sawmills and pulp and paper mills), and are therefore directly exposed to international market pressures. Hence, the requirements of publishing houses and buyer groups meant that FSC certification translated into a competitive advantage for the forest companies. By contrast, among the Swedish forest owner associations, only Södra operates its own sawmills and pulp mills. Participation in the Swedish FSC working group was pushed by Södra, representing private landowners in the south, whereas the decision to leave was strongly advocated by the private landowners in the north. Södra was, in fact, inclined to remain on the FSC working group, even without participation from the other associations (Elliott 1999, p. 387), but in the end, it joined the other associations when they abandoned the working group in 1997. Because of increasing market demand, however, Södra now offers FSC certification in addition to PEFC certification to its more than 50000 members in the south of Sweden.

Unlike Sweden, Norway has only one major forest company – Norske Skog – which, since its establishment by the forest owner associations in 1962, has become one of the world's largest pulp and paper companies. Most Norwegian pulpwood is sold to Norske Skog. As in Sweden, the development of forest certification standards in Norway was largely a response to market demands in Germany, the UK and other European markets. Although Norske Skog is a major pulp and paper company, it is a minor forest owner, particularly compared to the Swedish forest companies. In fact, the company's strategy in recent years has been to sell off its forests and specialize in the processing of printing paper, with the result that it now owns little forestland (Gulbrandsen 2003). The choice of certification scheme was thus left to the forest owner associations, primarily representing owners of small forests.[12] Although FSC certification would have been a benefit for Norske Skog in European export markets, the company did not pressure the forest owners to choose this program. Traditionally, there have been close ties between Norske Skog and the forest owners, who, despite their loss of influence as the company has expanded over the past two decades, are still among the largest shareholders in the company. This relationship helps to explain why Norske Skog did not require FSC certification from its timber suppliers.[13] As noted, FSC certification would require further elaboration of the Living Forests standards through a multi-stakeholder process, but this would not be necessary with ISO 14001. Hence, the choice of ISO over FSC may be regarded as a utility-maximizing choice, in the sense that forest owners could ensure, at the lowest cost, that Norske Skog would continue buying their timber and the company's customers would continue buying its printing paper (Gulbrandsen 2003).

The development of FSC-competitor schemes in Norway and Sweden – driven by forest landowners – resembles steps taken in many countries. Small landowners believed that they had little say in FSC and that the scheme was made for industrial forest companies. A certification scheme's legitimacy among target groups and its sensitivity to their needs thus appear to be important for their participation. Because Norwegian forest owners distrusted FSC and believed that environmental interests and forest companies dominated the decision-making processes, it has been extremely difficult to sell this scheme in Norway. The fact that environmental interests, in collaboration with the industrial forest companies, were at the forefront in establishing FSC in Sweden contributed to the small forest owners' dissatisfaction with the process and their rejection of the outcome. What the forest owners feared most was that FSC would obtain a monopoly on forest certification, which, in their view, would mean that environmentalists, social groups and industrial companies in partnership could dictate the terms for forest management.[14] The formation of PEFC may be seen as a strategic move to regain control over an area dominated by environmental interests.

Unlike many certification processes elsewhere, however, the emergence of a forest owner-based certification system in Sweden did not marginalize

FSC there. On the contrary, FSC retains a stronger position in Sweden than in most other countries – a success that can be partly explained by the leadership role of environmental NGOs in promoting the scheme, but more significantly by the presence of a group of industrial forest companies, with economies of scale, large-scale forestry operations and organizational resources enabling them to handle FSC certification.

As Cashore et al. (2004) have explained, the corporatist tradition and the strong associational systems in the forest sector accelerated both support and opposition toward FSC in Sweden. The existence of well-organized landowner associations facilitated collective efforts to create a landowner-dominated program in response to supply-chain pressure to adopt FSC. The fast and widespread adoption of such a program among the nonindustrial forest owners in Norway can also be explained by the strong position of these associations. Similar associational features among the Swedish forest companies worked to accelerate support for FSC, as members of the Swedish Forest Industries Federation fell in line with their association.

In summary, variation in forest industry structure goes a long way toward explaining divergent certification choices in Sweden and Norway. The Swedish forest companies responded to market pressures and opportunities by choosing the widely recognized FSC scheme, whereas the lower market exposure of nonindustrial forest owners in both Sweden and Norway meant that they could afford to reject FSC and develop a competing program.

Mutual Recognition Discussions in Sweden

From the industry's point of view, the main purpose of certification is to improve, or at least maintain, market access for certified organizations and labeled products. The so-called chain-of-custody tracking tells consumers that products carrying the FSC eco-label come from a certified forest. The chain-of-custody certificate allows forest products to carry the FSC label if a certain percentage of the wood, chip or fiber contained in those products can be traced back to FSC-certified forests. Initially, FSC required solid wood products bearing its label to contain 100 percent certified wood; and chip, fiber and component products to have at least 70 percent FSC-certified content. Were supplies of FSC timber to pulp and paper mills and sawmills to fall below these thresholds, none of the output could carry the label. The Swedish forest companies had problems meeting these strict labeling requirements. Deliveries of PEFC-certified and non-certified timber were particularly difficult for those mills least self-sufficient in FSC wood. Because they were unable to meet chain-of-custody requirements, they could not market their wood products under the FSC logo. The problem was exacerbated by the Swedish 'wood swapping' system, whereby harvested timber is sent to the nearest mills – regardless of ownership – in order to reduce transportation costs (Cashore et al. 2004, p. 206).

The competing PEFC scheme offered the forest industry a better deal with its chain-of-custody certificate based on a 'percentage-in, percentage-out' labeling approach, which operated with a given percentage of PEFC material for each production batch instead of imposing absolute thresholds (Cashore et al. 2004, p. 210). If 30 percent of the wood, chip and fiber supplied to a paper mill was certified, for example, then the mill could sell 30 percent of its paper products as certified. In 2000, in order to ward off competition from PEFC and attend to industry needs, FSC reduced labeling thresholds (FSC 2000). Yet, many mills still found it difficult to obtain enough FSC timber to meet the lower targets. As a result, FSC-certified volumes were too small to meet the demand from buyer groups, such as WWF's group in the UK.[15]

Mutual recognition by FSC and PEFC – the reciprocal recognition of the schemes in terms of purpose, process and outcome – would resolve the problem for the forest industry. Combining the two timber flows from FSC and PEFC would provide the mills with large volumes of certified timber, thereby enabling them to meet the demand for FSC-certified wood. Because mutual recognition at the international level was not a likely prospect, the Swedish forest industry invited the environmental NGOs to participate in the process of working out a mutual recognition framework or building a bridge between FSC and PEFC in Sweden called 'the Stockdove process'. Their aim was to create a 'Swedish certified' system, which would enable them to sell all timber sourced from certified operations in Sweden as FSC-certified, gaining the ultimate market recognition.

The forest industry and environmental NGOs agreed to appoint ecologist Gustaf Aulén from the forest association, Södra, and ecologist Stefan Bleckert from WWF to examine the certification documents and assess the differences. The side-by-side comparison revealed that PEFC would have to strengthen 17 of its standard elements to reach compatibility with FSC, whereas FSC would have to strengthen only four standard elements to reach compatibility with PEFC (Aulén and Bleckert 2001). The Swedish PEFC immediately moved to put a standard revision process in motion, with the aim of adopting FSC-like standards (Cashore et al. 2004, p. 212). As discussed, the scheme was subsequently adjusted upward to match FSC's level of environmental stringency. Yet, the process did not result in mutual recognition, because the Swedish environmental NGOs did not approve of the governance and operation of PEFC. They claimed that PEFC gave forestry interests ultimate control over rulemaking and interpretation, and, as noted, they refused to participate in or support the scheme as long as environmental and social stakeholders had limited decision-making influence. Moreover, the investment that environmental NGOs have made in FSC renders it highly unlikely that they would ever endorse a competing program. Their primary motivation for supporting the Stockdove process and partaking in 'official' side-by-side comparison of FSC and PEFC was to prove their point that FSC was the stronger standard.[16]

In an effort to circumvent the problems emanating from the existence of two competing schemes, four of the FSC-certified Swedish forest companies pursued PEFC-style certification. By pursuing parallel certification, they moved strategically to adapt to market demands for bigger volumes of certified timber. The forest companies also lobbied FSC to allow greater flexibility in chain-of-custody tracking.[17] In 2004, in response to the lobby, FSC drafted new and more flexible chain-of-custody standards for companies supplying and manufacturing FSC-certified products (ENDS 2004b). Mutual recognition of FSC and PEFC in Sweden does not seem likely, however. It would erode FSC's market advantage and position as the only forest certification scheme supported by a wide range of environmental and social NGOs.

THE ON-THE-GROUND IMPACT OF CERTIFICATION

At the end of the day, forest certification will be judged on its ability to change forestry practises in ways that reverse the environmental deterioration of forests. This is partly a question of the stringency of certification standards and partly a question of compliance with those standards. Compliance with standards can be expected to be encouraged by the authority of certification bodies to audit forest operations, address issues that must be improved and suspend the certificate of noncompliant forest owners.

Certification, it appears, is vital to achieving public policy goals in the Swedish forest sector. The Swedish Parliament's 1999 objective to see a further half million hectares of high-conservation-value forestland preserved on a voluntary basis by the year 2010 was already exceeded in 2004 through the implementation of forest certification standards.[18] In 1999, in addition to the conservation target, Parliament adopted targets for expanding total land resources containing dead wood, mature forests with deciduous-dominated stands and old growth forest, in order to enhance biological diversity. These objectives were (1) to increase the quantity of hard dead wood by at least 40 percent throughout the country and considerably more in areas where biological diversity is particularly at risk; (2) to increase the area of mature forests with a large deciduous element by at least 10 percent; (3) to increase the area of old forest by at least 5 percent; and (4) to increase the area regenerated with deciduous forests (SEPA, 2004, pp. 65–6). Success in this area depends largely on the way productive forests are managed, including the stands that are chosen for felling and the degree to which large, dead trees are retained (SEPA 2004, p. 66). According to the Swedish Environmental Protection Agency (SEPA), the objectives are likely to be achieved by the target year 2010 (SEPA 2004, p. 66). Although state agencies have initiated such educational campaigns as 'Greener Forests' to improve forest management, SEPA and the National Board of Forestry agree that certification is the most important initiative in the forest sector to improve management practices.[19]

A study commissioned by WWF and the Swedish Society for Nature Conservation showed compliance with many FSC requirements in forest operations in Sweden, but also that some rules had been only partially implemented (Dahl 2001). The study found that forest owners complied with requirements concerning key habitats, forests exempted for nature conservation, transition zones and buffer zones, soil scarification and natural regeneration. Partially implemented requirements included those addressing trees with high biodiversity value; trees with a good chance of developing into large, old trees; dead wood; landscape ecology planning; balanced age distribution in a landscape perspective; and red-listed species outside key habitats. Two-thirds of more than 400 Corrective Action Requirements (CARs) issued by certification bodies between 1996 and 2001 addressed ecological issues, a quarter concerned social issues and less than 2 percent addressed economic concerns such as productivity and yield (Dahl 2001). This distribution of CARs indicates that implementing environmental standards is the most serious challenge to forest companies, but also that certifiers are ready to address breaches of those standards.

Environmental NGOs claim that third-party auditing in PEFC and other forest-owner-dominated schemes generally suffer from flexible standards, lack of transparency and leniency in the assessments of compliance (Vallejo and Hauselmann 2001; Dahl 2002; Ozinga 2004). Although FSC, too, has been criticized for deficient auditing procedures (for example Counsell and Loraas 2002), the stringency of its standards may facilitate credible verification of compliance (Gulbrandsen 2004). Moreover, FSC-accredited certifiers issue public summaries of certification and audit reports, and their auditing procedures are generally transparent and open to stakeholder participation. By contrast, auditing reports have often been difficult to obtain from PEFC, and there is no public registry of the scheme's private forest owner members. This lack of transparency in the PEFC system has made it difficult to scrutinize and assess the impact of third-party auditing by examining the distribution of CARs (Dahl 2002). Even so, it is possible to assess the impact of certification by examining environmental indicators on forest management units. A study comparing the impact of certification in southern Sweden indicates that FSC-certified owners set aside more forestland and leave more dead wood and deciduous-dominated stands than PEFC-certified owners do (Andersson 2002). The study examined only ten FSC-certified and ten PEFC-certified private forest management units, however, and is therefore relatively limited.

As of mid-2009, only one study had been published ascertaining the impact of certification on forest biodiversity in Norway. Sverdrup-Thygeson et al. (2008) investigated several environmental indicators on 236 forest regeneration areas (118 before certification and 118 after certification) in southeastern Norway. Their study showed an increase in the number of retention trees and an increasing mean width of buffer strips left along rivers and lakes on certified units, but demonstrated that more retention trees must be left in order to comply with certification standards. The limited number of environmental issues and the limited geographical area investigated make it difficult to draw any general conclusions about the impact of certification on the protection of forest biodiversity.

Although research indicates that forest certification in general has resulted in improvements in internal auditing and monitoring among forest organizations (see Chapter 4), we still know too little about the environmental impact of forest certification and its efficacy as a problem-solving instrument. These are areas in urgent need of closer examination.

BROADER CONSEQUENCES: SHIFTING ALLIANCES AND NEW CLEAVAGES?

Certification processes could facilitate dialogue among various stakeholders and conflict resolution over forestry practices, but they could also result in shifting alliances and new cleavages. In applying the *advocacy coalition framework* (Sabatier 1998) to the Swedish forest certification process, Elliott and Schlaepfer (2001) argue that the process allowed policy-oriented learning in the FSC working group and contributed to changes in the 'core beliefs' of participants. Prior to the certification process, they argue, there were two opposing coalitions in Sweden: an environmental coalition and a forestry coalition. Because of the certification process, the argument goes, the two groups merged into a 'sustainable forestry coalition' comprising all the organizations that agreed upon the 1998 FSC standard. The nonindustrial forest owners became marginalized as a result of their decision to leave the FSC working group. Based on the advocacy coalition framework, we should expect forest companies and NGOs to continue to work together, and the nonindustrial forest owners to become targets of environmental group campaigns.

There is little doubt that forest certification contributed to better relations between the forest industry and NGOs in Sweden, particularly during the work in the FSC working group and the first years after agreement on the Swedish standard.[20] In fact, the large-scale FSC implementation in Sweden became the much-needed success story of WWF and other environmental NGOs, which had invested a great deal of their resources and prestige in promoting the scheme in the marketplace. On the other hand, environmental pressure groups are in danger of losing much of their influence when certification processes are accepted and conflict levels abate. There is a risk that they will be satisfied with what has been achieved and fail to act should new knowledge or evidence of harmful forestry practices appear. Whether an 'advocacy coalition' composed of the Swedish forest companies, environmental NGOs and the other participants in FSC actually exists is less clear. Following a few peaceful years, antagonism within the forest sector seems to be on the rise again, albeit still far below the conflict levels of the 1970s and 1980s. Forest companies have crossed swords with activists and local communities over logging, and the media have reported violations of FSC standards. In the opinion of the forest industry, many of these conflicts are associated with differing expectations about what the certification processes would deliver. The forest companies believed that certification would ensure trust in their claim that Swedish forests were well managed even before the certification processes began. By contrast, many environmentalists expected a radical shift in forestry practices as a result of certification. Consequently, the environmentalists were disappointed over the lack of large-scale changes in forest management following certification. There has been some concern within the environmental movement that by collaborating with the forest industry in FSC, they are being co-opted by a neoliberal, market-based approach to environmental governance, stifling their ability to advocate for radical changes in forest management. According to some environmental groups, noncompliance with certification requirements also looms large. A forestry network within the Swedish Society for Nature Conservation has reported many instances of alleged breaches of FSC standards, and the network has published on the Internet almost 400 instances of what it calls irresponsible logging.[21] Such disputes show that certification is not a panacea, but also that certification rules may be used by NGOs in an attempt to hold forest companies accountable.

In summary, the evidence does not support the idea, based on the advocacy coalition framework, that a 'sustainable forestry coalition' emerged as a result of forest certification. Traditional forestry conflicts, such as those concerning protection of primary forests, key habitats and forests of recreational value, still tend toward arguments between forest companies or landowners on the one hand and environmentalists or local communities on the other. FSC collaboration necessarily disciplines the parties, however, obliging them to work together to find compromise solutions on contentious issues. In addition, repeated interaction over time in organized networks can build mutual trust, common expectations of what is right and proper conduct, and internalization of norms (Cutler et al. 1999; Boström 2006a). The participants in FSC represent very different constituencies, but all interviewees attest to increasing understanding and appreciation of the varied interests in forestry. Indeed, most participants say that it has enabled dialogue in a conflict-ridden sector and brought them closer together. Although post-certification conflicts shed doubt on claims that 'core beliefs' have changed among participants, they have learned to respect each other's competencies and ideas and, by establishing a permanent Swedish FSC Council, they have created an institution for deliberation and conflict resolution.

CONCLUSIONS

This chapter has shown that variation in forest ownership and export dependence was particularly important for decisions to adopt FSC or a landowner-dominated program. The Swedish forest sector comprises large companies and small, nonindustrial forest owners, whereas the Norwegian forestland is controlled almost entirely by nonindustrial forest owners. The Swedish forest companies, controlling almost 40 percent of the forestland in Sweden, opted for FSC to increase or at least ensure market access in Germany, the UK and other export markets. They were directly exposed to market pressures and NGO targeting and were consequently susceptible to supply-chain demands to adopt FSC. The nonindustrial forest owners in Norway and Sweden rejected FSC because of its relatively stringent environmental and social standards and what they regarded as its lack of legitimacy and its inflexibility in accommodating the needs of small-scale, nonindustrial forestry. The forest owners feared that FSC would obtain a monopoly on forest certification, which, they believed, would leave them with little influence over forest management. And because they were less exposed to market demands than the forest companies were, they could afford to reject this program. Their strong associational systems facilitated the creation of landowner-dominated programs in response to supply chain pressure to certify.

One robust finding consistent with the work of Cashore et al. (2004) emerges from this comparison of forest certification in Sweden and Norway: export-dependent forest companies are more likely to adopt FSC than are those selling primarily in the domestic market. Another result, also consistent with the observations of Cashore et al. (2004), is the finding that large, vertically integrated forest companies are more likely to adopt FSC than are small, nonindustrial forest owners. Yet, these variables do not fully determine certification choices. Other factors could trump export dependence and convince forest companies to adopt an FSC-competitor program. The forest companies in Finland, for example, comparable to the Swedish forest companies in size and export dependence, rejected FSC and adopted PEFC instead. Explanations have indicated widespread opposition toward FSC among the hundreds of thousands of Finnish forest owners, who could more easily influence certification processes because of their importance as a supplier of certain fine paper grades (Cashore et al. 2007b). Similarly, the fact that the Swedish forest companies were dependent on export markets did not stop them from considering FSC-competitor programs, although they adopted FSC in the end. The broader point is that the basis of certification choices is complex and influenced by a number of factors pulling in different directions. One such factor, often overlooked in the literature, is the bargaining processes that occur within standard-setting groups. In the processes investigated in this chapter, we see that strategic choices at critical junctures created 'lock-in' effects (Pierson 1993) that limited future choices and significantly increased the cost of changing course (cf. Cashore et al. 2004). The decisions of the forest owner associations to leave the Swedish FSC working group, for example, represented a key decision point, which fundamentally changed the course of forest certification in Sweden. The forest owners were, in fact, close to accepting the proposed FSC standard, but they withdrew, largely over disagreement with the Sami representatives about reindeer herding on private forestland, choosing, instead, to forge a landowner-dominated program. If they had decided to remain within the working group and continue the negotiations, there may have been only FSC-certified forestland in Sweden today rather than two competing schemes.

Regarding the impact of forest certification, it is vital to its credibility as an environmental standard to be able to demonstrate that it is making a difference in on-the-ground practices. What can be said about the true problem-solving ability of forest certification? A common assumption is that the stronger the environmental performance standards, the greater the impact on forestry practices. Although this relationship might have existed in a comparison of the relative improvement of forest holdings, little consistent support was found at the aggregate level. We have seen that not all target groups will accept a scheme with stringent and rigorous standards, which obviously foreshortens the scheme's ability to change widespread forestry practices. In Norway, Sweden and most countries in which FSC is established, less intrusive and more discretionary schemes have emerged.

In conclusion, our limited knowledge of the genuine problem-solving ability of forest certification remains a major constraint. Although forest certification seems to have modified on-the-ground practices in ways that lead to less environmental deterioration of forests, we still know too little about the environmental impact of forest certification. Future research should investigate the relative environmental improvements of certified forest holdings and the impact of certification on forest biodiversity.

NOTES

1. Statistics Norway: www.ssb.no/english/subjects/10/04/20/skog_en/ (accessed June 25, 2009).
2. Interviews, Swedish Forest Industries Federation, October 13, 2004, Sveaskog, October 14, 2004 and SCA, November 26, 2004.
3. Interviews, WWF Sweden and Swedish Society for Nature Conservation, October 14, 2004.
4. Interview with former senior ecologist in AssiDomän (now a subsidiary of the state-owned Sveaskog forest company), October 22, 2004.
5. Interview, October 13, 2004.
6. Interview, Swedish Federation of Forest Owners, October 15, 2004.
7. Interviews, The Norwegian Forest Owners Federation, April 10, 2004, and Norske Skog, October 5, 2004.
8. Interview with senior advisor, Ministry of Agriculture, June 26, 2001.
9. Interview, WWF Norway, September 25, 2003.
10. The table is based on the Swedish FSC standard (FSC Sweden 1998) and the Norwegian Living Forests standard (Living Forests 1998).
11. Interview, The Norwegian Forest Owners' Federation, April 10, 2003.
12. Interview, Norske Skog, October 5, 2004.
13. Interview, Norske Skog, October 5, 2004.
14. Interview, The Norwegian Forest Owners' Federation, April 10, 2004.
15. Interview, Sveaskog, October 14, 2004, and SCA, November 26, 2004.
16. Interviews, WWF Sweden, and Swedish Society for Nature Conservation, October 14, 2004.
17. Interviews, Swedish Forest Industries Federation, October 13, 2004 and SCA, November 26, 2004.
18. As early as 2002, approximately 990 000 hectares of forestland had been set aside voluntarily (SEPA, 2004, p. 65).
19. Interviews, SEPA, October 11, 2004; and National Board of Forestry, October 13, 2004.
20. See also Hellström (2001).
21. Available in Swedish at: www.snf.se/pdf/dok-skog-exempelsamling.pdf.

6. Spillover to the fisheries sector: the Marine Stewardship Council

This chapter examines how the certification model was exported from the forest sector to the fisheries sector, detailing the origins and evolution of MSC – the first and only global certification program for wild-capture fisheries. More than any other certification initiative, the creation of MSC was inspired by FSC's success in the forest sector. Particular attention is paid to the influence of choices of program features on the development of the program. The chapter demonstrates that although MSC mimicked FSC, the initiators of the program purposefully avoided three of FSC's key features: (1) they decided against an open-membership organization; (2) they chose to avoid delegating authority to make a global standard locally appropriate to national affiliates; (3) they decided not to address social issues in the principles and criteria of the program. These decisions were pivotal for the type of debates that emerged about fisheries certification and the subsequent development of MSC.

The chapter begins with a brief review of challenges to fisheries management and the evolution of intergovernmental governance through multilateral and regional fishing agreements. The second section is an examination of the emergence of single-species labels and seafood buyer guides as consumer-based responses to specific environmental or management problems in the fisheries sector. The third section details the formation and features of MSC, including its governance structure, certification standards and certification process. The fourth section examines how governments and certain producers responded to the emergence of MSC, paying particular attention to the development of international guidelines for eco-labeling. In the fifth section, the influence of stakeholders in the program is discussed, with particular emphasis on one question: Does MSC empower those who are marginalized in traditional fisheries governance? The conclusion reflects upon the question of how choices of program features, scope and governance influenced debates about fisheries certification and the development of MSC.

INTERGOVERNMENTAL FISHERIES GOVERNANCE

Coastal and open-ocean fish stocks are common-pool resources; they are nonexclusive and scarce in supply (Ostrom 1990). In principle, anyone can fish in the ocean and no one can exclude other people from fishing, but because fish have limited reproductive capacity, one person's use of the resource has consequences for others. Accordingly, in order to avoid a 'tragedy of the commons' (Hardin 1968), there must be management regimes that limit the number of users or otherwise restrict the capture of the resource (Hoel and Kvalvik 2006).

The global imbalance between fish resources and harvesting capacity is the most serious environmental problem in the fisheries sector. Driven by the rapid growth of the world's fishing fleet, world marine capture fisheries production quadrupled between 1950 and 1990 (FAO 2002). Although it has since leveled off or even declined, there is talk of a global crisis in marine capture fisheries (Watson and Pauly 2001). Decades of overfishing – largely a result of overcapacity in the fishing fleet and lack of effective regulations to restrict access and resolve distributional issues – have had dire consequences. According to the latest *State of World Fisheries and Aquaculture Report* by the Food and Agriculture Organization (FAO) of the United Nations (UN), more than one-quarter of the fish stock groups monitored by the FAO were either overexploited (19 percent), depleted (8 percent) or recovering from depletion (1 percent) (FAO 2009, p. 7). The proportion of overexploited and depleted stocks increased from about 10 percent in the mid-1970s to around 25 percent in the early 1990s, where it has stabilized until the present (FAO 2007, p. 29). Overfishing not only threatens the reproductive capacity of a fish stock; it also affects ecosystems through by-catches of nontarget species, habitat destruction and vessel pollution. Intensive fisheries methods, such as bottom-trawling, devastate habitats, and heavy fishing for a target species by methods such as purse-seining (fishing with a long vertical net) often results in significant capture of nontarget species. The high incidence and increasing sophistication of illegal, unreported and unregulated (IUU) fishing adds to the problem of overfishing. Widespread use of flags and ports of convenience exacerbates the scope and extent of IUU fishing (FAO 2007, pp. 71–7).

Governments have recognized that as long as open access to marine fish stocks is maintained, fish stocks are likely to remain overexploited. They have therefore created elaborate international and regional regimes for the management and conservation of living marine resources. These regimes are centered on the 1982 UN Law of the Sea Convention, which entered into force in 1994, and provides a legally binding framework for regulating all use of the world's oceans (Orrego Vicuña 2001). The Convention codified the right of coastal states to create 200-nautical-mile Exclusive Economic Zones (EEZs) where they have sovereign jurisdiction over the natural resources. According to the Convention, coastal states must manage fisheries in a sustainable manner, cooperate with international organizations toward this end and promote the optimum utilization of fisheries resources (Articles 61 and 62). Coastal states are required to cooperate to ensure the conservation and development of stocks when the same stock or stocks of associated species occurs within the EEZs of two or more states (Article 63). In short, the extension of national jurisdiction over larger parts of the world's oceans meant that stocks that used to require multinational management arrangements could now be managed by one state or through the collaboration of a few states (Orrego Vicuña 2001). Management requirements for straddling and highly migratory stocks, such as tuna, remained less clear, and many coastal states continued to expand their fishing fleets and were lenient in implementing the commitments of the Law of the Sea Convention (Stokke 2001).

The institutional response to the inadequate regulation of high-seas fishing was the 1995 creation of the UN Fish Stocks Agreement[1] (deFontaubert 1995), which entered into force in 2001. It strengthened the obligation for regional cooperation among coastal states to promote conservation of straddling and highly migratory fish stocks. Introducing a precautionary approach to fisheries management, the Fish Stocks Agreement established requirements for high-seas fishing, rules for effective enforcement and procedures for mandatory dispute resolution (Balton 1996). Another instrument adopted in 1995 was the FAO Code of Conduct for Responsible Fisheries (FAO 1995). It was issued by FAO to establish principles and guidelines for responsible fishing practices with a view to ensuring the effective conservation, management and development of fisheries around the world. Serving as a comprehensive guide to appropriate fisheries management, it comprises a number of nonbinding guidelines, some of which are binding under other agreements. The so-called Compliance Agreement,[2] for example – a legally binding treaty to strengthen the obligations of flag states – is considered part of the Code.

A number of International Plans of Action and technical guidelines have been adopted under the FAO Committee of Fisheries to implement the Code of Conduct, elaborating on issues such as by-catches, the overcapacity problem and IUU fishing. At the 2002 World Summit on Sustainable Development, states agreed that by 2010 an ecosystem-based approach to fisheries management must be implemented, and that by 2015 overfished stocks should be rebuilt. In spite of the evolving institutional framework for fisheries management, however, overcapacity in the fishing fleets, by-catch of nontarget species, harmful fishing methods and IUU fishing by, for instance, vessels under flags of convenience, remain serious problems in the fisheries sector (FAO 2009). It was the belief of many conservation organizations that government regulations had failed to resolve the problems in the fisheries sector. Although fisheries in some regions are relatively well-regulated, regulations are absent or poorly enforced in many regions of the world. Seafood labeling, buyer guides and certification schemes thus emerged in response to the fisheries management challenges. Rather than relying on government regulations, these schemes are based on harnessing consumer power to encourage behavioral change in the fisheries sector. In the next section, the formation of the first seafood labeling initiatives in the fisheries sector is examined.

SINGLE-SPECIES ECO-LABELS AND SEAFOOD BUYER GUIDES

Social movement activism and consumer concern were key drivers behind the first eco-labeling initiatives in the fisheries sector. The inadvertent capture of nontarget species (by-catch) such as marine mammals and sea turtles is a serious problem in fisheries management, but it can be resolved or alleviated by adopting special fishing gear and methods. Mounting public concern over the substantial dolphin by-catch by tuna fisheries helped to prompt the formation of the first dolphin-safe labeling scheme by the Earth Island Institute, a US-based conservation organization (Teisl et al. 2002). Use of the label, introduced in 1990, indicates that fishers do not use fishing gear in the catching of tuna that results in the by-catch of dolphins. The US government created its own dolphin-safe label under the Dolphin Protection Consumer Information Act (1990), which established rules for the tuna catch and the labeling of tuna products. Another program was introduced in 2001, when the Inter-American Tropical Tuna Commission (IATTC) supplemented restrictions on fishing to reduce dolphin by-catch with a certification procedure and an eco-label to mark tuna that were caught by IATTC member countries and vessels (Ward 2008). In order to reduce sea-turtle mortality, single-species labeling schemes were also introduced, by issuing a turtle-safe label on shrimp products to guarantee that the fishing method would not kill sea turtles.

The development of seafood buyer guides was another consumer-based approach that was introduced in the hope of improving fisheries governance and helping consumers to chose fish from sustainably managed fisheries. The Audubon Seafood wallet card, which was intended to guide customers in ordering seafood in restaurants and buying fish in supermarkets, identified seafood choices according to traffic-light colors: green, yellow and red. This card is no longer updated,[3] however, and consumers looking for it on Audubon Society webpages are encouraged to download the latest seafood guide from Monterey Bay Aquarium, which distributes and updates Seafood Watch guides for consumers.[4] Anticipating visitor questions about making better seafood choices, Seafood Watch began when Monterey Bay Aquarium developed a list of sustainable seafood as part of their 1997–1999 Fishing for Solutions exhibit. The Seafood Watch wallet card identifies species that are abundant, well-managed and fished or farmed in environmentally friendly ways ('best choices'), seafood that are acceptable alternatives to the best choices ('good alternatives'), and seafood that are overfished and/or fished or farmed in ways that harm other marine life or the environment ('avoid'). In addition to a national guide, Seafood Watch creates regional guides that contain the latest information on sustainable seafood choices available in different regions of the USA. Yet another initiative was introduced in 2001, when Sea Web, a US-based nonprofit foundation, launched its Seafood Choice Alliance program, in the hope of building a larger market for ocean-friendly seafood. This program followed on the heels of the 'Give Swordfish a Break' campaign, a further example of the many single-species campaigns launched in the late 1990s and early 2000s.

In spite of the merits of the single-species approach, it soon became clear that concentration on a single facet of environmental protection did little to address major environmental problems in the fisheries sector. The significant attention to individual species like dolphins and sea turtles seemed even to slow the development of a sector-wide approach to certifying sustainably managed wild-capture fisheries (Auld 2007). In these cases, the problems addressed by the labeling schemes were animal rights and the protection of endangered species – not such problems as overfishing, depletion of fish stocks and adverse marine ecosystem effects (Allison 2001, p. 945). The strong focus on dolphin and sea turtle protection, particularly in the USA, might thus have diverted consumer attention away from the dire resource management challenges in the fisheries sector.

On the other hand, the distribution of buyer guides might have increased consumer awareness about overfishing and depleting fish stocks, thus preparing the market for a sector-wide certification scheme. In any event, buyer guides, unlike certification schemes, do not involve standard setting for sustainable fishing practices and third-party inspections of fisheries, a situation that raises questions about their effectiveness. A study commissioned by the Monterey Bay Aquarium concluded that the distribution of more than one million Seafood Watch wallet cards had neither brought about changes in the seafood market nor reversed the decline of targeted fish stocks (Seafood Watch Evaluation 2004). Yet, Monterey Bay Aquarium continues to promote Seafood Watch, hoping that downloadable versions and new applications such as 'Seafood Watch Recommendations on your iPhone' will bring about changes in consumer behavior.

The effectiveness of seafood awareness campaigns has also been questioned because of the widespread renaming and mislabeling of fish species in the seafood market. Some fish species are given more appetizing-sounding names in order to increase sales, and others are mislabeled as different species in the hope of concealing illegal or unsustainable fishing. In the USA, where 80 percent of seafood is imported, more than one-third of all fish is mislabeled, a situation that can easily mislead concerned, albeit uninformed consumers into purchasing endangered or overfished species (Jacquet and Pauly 2008).

THE FORMATION AND EVOLUTION OF MSC

Although some environmentalists wanted the seafood-ranking approach to be the way of sensitizing consumers about their purchasing practices, others believed that only a sector-wide certification system, akin to FSC, could address major environmental problems in the fisheries sector. The success of FSC's sector-wide approach served as a major motivation for the WWF to develop a similar certification scheme in the fisheries sector. Beginning with the origins of the program, this section examines the development of MSC's governance structure, its principles and criteria and its certification process.

From a WWF-Unilever Partnership to a Fully Independent Organization

Whereas FSC emerged from a broad coalition of non-governmental groups, MSC began as an NGO-business partnership between WWF and the multinational corporation Unilever, the world's largest buyer of seafood at the time. WWF, wanting a partner able to facilitate the uptake of certified products among supermarket chains and other retailers, initiated the partnership. As a major player in the fish food sector with a respectable sustainability policy, Unilever fitted the bill. As argued by Michal Sutton, then Director of WWF's Endangered Seas Campaign, the idea was to harness market forces to encourage behavioral change in fisheries: 'When Unilever and other major seafood companies make commitments to buy their fish products only from well-managed and MSC-certified fisheries, the fishing industry will be compelled to modify its current practices. Governments, laws and treaties aside, the market itself will begin to determine the means of fish production' (Sutton 1996, p. 18).

The initiation of MSC was inspired by FSC's success, and the similarity of their names and logos was no coincidence. Murphy and Bendell (1997) describe how staff members engaged in WWF's Endangered Seas Campaign learned informally of the FSC certification model from their colleagues, considered its application to fisheries, and decided to create a similar model for fisheries certification. The fact that a member of WWF's forest team was contracted to investigate the possibility of creating a 'fisheries stewardship council' is further evidence of FSC's influence on early work toward establishing a similar scheme in the fisheries sector (Murphy and Bendell 1997). Unilever had also witnessed the achievements of FSC certification in the forest sector. Antony Burgmans, then responsible for Unilever's frozen fish business and later CEO of the company, recalls that for him and his colleagues, admiration for the work that WWF had done in the establishment of FSC motivated them to start discussions with the conservation organization about developing a similar scheme for fisheries (Burgmans 2003). The idea of a partnership with WWF to create a fisheries certification scheme was approved within Unilever largely because it won Burgmans' support (Fowler and Heap 2000, p. 139). Being the world's largest purchaser of frozen fish, Unilever was under considerable pressure from environmental NGOs to address the environmental problems in the fisheries sector. Greenpeace in particular was campaigning against Unilever to pressure the corporation to adopt environmentally sustainable purchasing policies for its fish brands.

The business-NGO partnership between WWF and Unilever was announced in August 1996. The two partners commissioned Coopers & Lybrand (now PricewaterhouseCoopers), a business consulting firm, to study how MSC could be governed. The consultants studied FSC's governance structures (Sutton and Whitfield 1996), and even attended its first General Assembly to learn from FSC's experience with establishing a forest certification scheme (Synnott 2005, p. 25). Based on this study, they advised against an open membership organization like FSC, in order to avoid cumbersome decision-making processes, complex structures and time-consuming procedures. This advice was accepted, and in 1997 WWF and Unilever established MSC as an independent nonprofit organization – without specific provisions for membership and without national affiliates to elaborate upon global standards and fit them to a local context. Although MSC did include national affiliates, they were solely developed for outreach and marketing – not for making global standards locally appropriate (Auld 2007). The authority to elaborate upon MSC's principles and criteria was instead given to the certification bodies that were to assess fisheries seeking endorsement by the program. WWF and Unilever wanted to prevent a governance arrangement that was inefficient, inflexible and expensive to operate. Yet there was controversy over MSC because of the major involvement of WWF and Unilever rather than a wide range of stakeholders in the initiation of the scheme. Several stakeholders wanted participation in a membership-based decision-making body like FSC's General Assembly (Constance and Bonanno 2000, p. 131).

MSC initially operated under an interim board comprising members from WWF and Unilever, and chaired by a senior partner at Coopers & Lybrand (Fowler and Heap 2000, p. 140), albeit in a personal capacity. Fishing communities in particular were dubious about the key role of WWF and Unilever in establishing MSC, as they had been since the inception of the scheme. As early as July 1996, a month before the official launching of the scheme, an editorial in *Samudra*, a periodical issued by the International Collective in Support of Fishworkers, commented that MSC 'has not won the total confidence of fishing communities, either in the South or the North, because of their great distrust of Unilever'. It added that the scheme 'would have been taken far more seriously by fish workers' organizations had WWF consulted them before plunging in' (*Samudra* 1996, p. 1). Several governments were also concerned about the implications of MSC for fisheries governance and fishers around the world (Gulbrandsen 2005b; Hoel 2006). As discussed later in this chapter, the formation of MSC provoked calls for intergovernmental rules for eco-labeling of fish products.

To fend off assertions that MSC was controlled by WWF and Unilever, MSC took several steps to strengthen its credibility and establish itself as a fully independent organization. In March 1998, an international Board of Trustees was established to oversee the scheme, with John Gummer, Member of Parliament and former UK government Minister for the Environment, accepting the position of Chair of the Board. Another step toward full independence was the withdrawal in June 1998 of seed funding from the two founding partners. Instead of being financed by WWF and Unilever, therefore, MSC had to raise funds from a range of private organizations, trusts and charities (Fowler and Heap 2000, p. 141). In fact, the transformation from a business-NGO partnership to a multi-stakeholder governance scheme, discussed below, was made possible by several grants from the Packard Foundation between 1999 and 2001, attesting to the tremendous importance of a few large foundations in supporting the program (see Chapter 7). By July 1998, MSC had become an independent nonprofit organization, which was seen by environmental organizations and the fishing industry alike as an essential first step in gaining credibility as a neutral body in a multi-stakeholder industry.

The Establishment of a Multi-stakeholder Governance Scheme

In addition to the Board of Trustees, MSC's founders had established a secretariat to run the day-to-day activities of the organization and coordinate the activities of the Advisory Board, a Standards Council and National Working Groups (Fowler and Heap 2000, p. 141). The Advisory Board resembled a membership body, but eligibility for participation remained unclear. In designing the structure of the Advisory Board, MSC's founders mimicked the three-chamber structure of FSC's General Assembly. The Advisory Board would include an economic chamber comprising fishers and other stakeholders that make a living from the seas; an environmental chamber composed of marine conservation groups and other NGOs; and a social chamber comprising stakeholders with an educational, social or consumer perspective on the use of marine resources. The key difference between this board and FSC's General Assembly was that rather than holding voting rights as in FSC, stakeholders represented in the membership-like body of MSC would have only an advisory role.

Notwithstanding the establishment of MSC as a nonprofit foundation, fully independent from WWF and Unilever, a range of stakeholders were concerned about unwieldy bureaucratic structures and a lack of transparent decision-making processes in the organization. In addition to the governance bodies mentioned here, MSC's governance structure included a number of committees and working groups with unclear rights and responsibilities, some of which never became operational. Indeed, the MSC governance structure had become 'excessively bureaucratic and complex, with so many rules, committees and requirements that some of the bodies attached to MSC had never met in person or even been appointed', whereas 'those bodies that did exist were appointed entirely by the MSC itself' (May et al. 2003, p. 33). This governance system was 'hardly a recipe for openness and transparency' (May et al. 2003, p. 33).

As a result of concerns such as these, MSC decided to change its governance structure. In 2001, following a ten-month governance review and consultation process, MSC announced a governance reform to enhance openness and responsiveness to various stakeholders within and outside the fisheries sector (MSC 2001a). The reform resulted in greater transparency, with the main international governance bodies now being the Board of Trustees, the Stakeholder Council and the Technical Advisory Board. In addition to these international bodies, National Working Groups were intended to provide guidance on MSC's activities and to promote the certification program. As noted, unlike national affiliates in FSC, they were not given the authority to make the global principles and criteria locally appropriate. This task was given instead to the certification bodies that appoint expert teams to assess fisheries seeking MSC endorsement.

The Board of Trustees is self-recruiting, and members are appointed – not elected – for three-year terms. It meets four times a year with a maximum of 15 members from industry, environmental organizations, the scientific community and seafood retailers.[5] As the highest decision-making authority in MSC, the board is responsible for ensuring that MSC meets its aims and is financially sound; approving and implementing the strategic direction of MSC; appointing new board members and key staff, including the MSC's chief executive; appointing members to the Technical Advisory Board; and publicly accounting for expenditures and income.

Replacing the Advisory Board, the Stakeholder Council can have a minimum of 30 members and a maximum of 50 members who meet annually to discuss MSC strategy, activities and other matters.[6] Its two joint chairs have seats on the Board of Trustees and are therefore involved in all board discussions and decisions. Members of the Stakeholder Council are drawn from (1) 'the public-interest category', composed of representatives from the scientific, environmental and marine-conservation communities (i.e. nonprofit constituencies) and (2) 'the commercial and socio-economic category', comprising individuals representing catch-sector interests; supply chain, processing and retail interests; and developing countries and fishing communities (i.e. profit-making constituencies). According to MSC's chief executive, the Stakeholder Council functions like a sounding board to help MSC develop, evolve and learn.[7]

Whereas the Stakeholder Council represents a wide range of interests pertaining to fisheries management and conservation, the Technical Advisory Board represents scientific and technical expertise, thus resembling the scientific advisory boards of many international environmental regimes. The 15 members of the Technical Advisory Board provide advice to the Board of Trustees on such technical matters as the development and application of principles and criteria. Being leading experts in certification procedures, fisheries science and ecological management, their role is to help MSC continually evolve policies and procedures. Before making final recommendations to the main board, the Technical Advisory Board will usually circulate proposals among stakeholder council members and other relevant groups. The chair elected by the Technical Advisory Board has a seat on the Board of Trustees that is similar to the two joint chairs of the Stakeholder Council. The representation of these chairs at the Board of Trustees bridges the previous gulf between the board and the range of constituencies with a stake in fisheries management (May et al. 2003).

The operations of MSC are run by a London-based international secretariat, staffed by some 30 people. Being responsible for day-to-day activities, including fundraising, accreditation, outreach and promotion, and developing and licensing the MSC logo, the secretariat has significant policy-making influence within the program. The secretariat is headed by a chief executive, who is appointed by the Board of Trustees. In addition to the international secretariat, MSC has established a number of regional offices for the purpose of outreach and promotion.

In summary, the governance reform resulted in an inclusive and transparent multi-stakeholder governance structure, with the Stakeholder Council, in particular, representing a wide range of nonprofit and profit-making constituencies. But the governance reform did not come with a similar power shift to empower environmental and social stakeholders. In order to avoid inefficient and time-consuming rulemaking processes, MSC left ultimate decision-making authority to the Board of Trustees rather than the Stakeholder Council. In this respect, MSC has upheld a more streamlined approach to stakeholder involvement than that of the membership-based FSC program.

The Comprehensiveness of the Principles and Criteria for Sustainable Fishing

MSC and FSC had different points of departure for standard development. Whereas FSC emerged partly in response to a failed set of intergovernmental negotiations on a forest convention, the MSC standards were based upon international fisheries agreements and recommendations. Building in particular on the FAO Code of Conduct for Responsible Fisheries, the principles and criteria of MSC were developed through an inclusive consultation process between 1996 and 1999. This consultation, involving more than 300 organizations and individuals, included two expert drafting sessions and a series of international workshops in various regions around the world. The work began in September 1996 with a meeting in Bagshot, UK, followed by workshops in Australia, New Zealand, Germany, the USA, Canada, Scandinavia and South Africa (Fowler and Heap 2000). The workshop in South Africa was the only one that took place in a developing country (Ponte 2008). Partly in response to criticism that MSC did not attend to the needs of fishworkers in the developing world (for example *Samudra* 1996; Kurien 1996), further consultations were held in a few developing countries (Sutton 1998; Fowler and Heap 2000).

As the work on the principles and indicators progressed, it became evident that MSC had to draw boundaries around what should and should not be included (O'Riordan 1997). In essence, MSC had to decide if the principles and criteria should address only fishing operations and environmental issues or if they should also address social and development issues (Auld 2007). Much of this debate occurred in the periodical of the International Collective in Support of Fishworkers, *Samudra*, and concerned the social aspects of fisheries management, particularly the needs of fishworkers and small-scale fisheries in developing countries (Auld 2007; Ponte 2008). Concerns were raised that MSC was not suitable for certifying fisheries in developing countries, given the many millions of fishworkers involved in small-scale fisheries, the numerous fish-landing centers and the diversity of species and fishing operations in the developing world. One commentator feared that the numerous small-scale decentralized fisheries in developing countries would be discriminated against, because they would be unable to bear the costs of certification and would not have the capacity to implement certification requirements (O'Riordan 1997). In rebuttal, Michael Sutton of WWF's Endangered Seas Campaign argued that certification under MSC could give southern fisheries a competitive edge over their northern counterparts, who had to contend with the collapse of many fish stocks in the North (Sutton 1998).

Whereas several commentators argued in favor of wide standards that encompassed both environmental and social issues (for example Mathew 1998), MSC decided to keep them narrower, focusing primarily on fishing operations and environmental issues in wild-capture fisheries. In this respect, MSC's approach differs from that of FSC, which focuses not only on environmental issues but also on social issues, including the rights of indigenous peoples, the long-term social and economic wellbeing of forest workers and local communities, and equitable use and sharing of benefits derived from the forests. By contrast, MSC does not address social issues such as fishing rights, the needs of fishworkers and the wellbeing of fishing communities. Its three main principles focus on the fishing activity and environmental issues, targeting specifically (1) the state of the target fish stocks, (2) the impact of the fishery on the ecosystem and (3) the performance of the fishery management system (see Box 6.1). These principles are supplemented by a number of more specific operational and management criteria. As discussed in the next section on the certification process, third-party certification bodies must elaborate on the principles and criteria to meet regional and local fishery conditions.

MSC also considered whether or not its standards should address fish farming. The significant growth of aquaculture production in the 1980s and 1990s had raised a number of environmental concerns, including the destruction of coastal habitats, nutrient and organic enrichment of recipient waters, negative population-level effects from escaped farmed organisms, eutrophication of lakes and coastal zones, veterinary drug

BOX 6.1: THE MSC PRINCIPLES

1. The state of the target fish stocks A fishery must be conducted in a manner that does not lead to overfishing or depletion of the exploited populations and, for those populations that are depleted, the fishery must be conducted in a manner that demonstrably leads to their recovery.

2. The impact of the fishery on the ecosystem Fishing operations should allow for the maintenance of the structure, productivity, function and diversity of the ecosystem (including habitat and associated dependent and ecologically related species) on which the fishery depends.

3. The performance of the fishery management system The fishery must be subject to an effective management system that respects local, national and international laws and standards and incorporates institutional and operational frameworks that require use of the resource to be responsible and sustainable.

Source: MSC (2002).

residues in aquaculture products and increasing demands on wild-capture fisheries for fishmeal to feed farmed species such as salmon (FAO 2007, p. 77). Large-scale shrimp farming in some areas has resulted in degradation of coastal mangroves and wetlands, and has caused water pollution. Expansion of commercial aquaculture has led to competition for space with fishers in some areas, threatening the viability of coastal small-scale fisheries. In addition, because farmed fish sometimes escape aquaculture pens, they could threaten the health and stability of wild populations. In some regions, escaped salmon have caused an increase in diseases and parasites among wild salmon. Beginning in the late 1980s, concerns such as these prompted organic certification organizations to introduce the organic labeling of approved fish-farming products such as salmon and shrimp. In 1989, the Soil Association, a UK-based organic certification organization, began work on a standard for organic farmed salmon and trout; the German-based certifier Naturland soon followed suit, releasing standards for organic pond farming in 1995 (Auld 2007). Selling organic farmed fish in supermarkets, however, could disadvantage non-labeled wild-caught fish. MSC sought to provide a label for wild-caught fish and decided, therefore, that its standards should not address aquaculture production. Yet the debate on whether or not MSC should expand into aquaculture certification has continued within the organization. A board statement issued in November 2006 indicated that MSC considered developing an aquaculture standard, but eventually decided to maintain its focus on wild-capture fisheries (MSC 2006a). This decision left evolving fish-farming-certification initiatives free of direct competition from MSC (Auld 2007).

The Certification Process

Whereas MSC is the standard setter, accredited certification bodies (certifiers) conduct the certification of applicants. The client for certification may be a fishers' association, an industry association representing quota holders, a processor's organization, a government management authority or any other stakeholder. The client in the certification of the Alaska salmon fisheries, for example, was the Alaska Department of Fish and Game (May et al. 2003, p. 19). A fishery must undergo a pre-assessment to determine if it can proceed to a full certification assessment. The pre-assessment is fully confidential, but clients sometimes release the outcome of the assessment on their websites to show stakeholders the identified needs for improvement.[8] When the fishery receives the result of the pre-assessment, it decides if it should move toward a full assessment. To ensure transparency, the fishery must publish when it goes into the full assessment – in a local newspaper, for example – and notify all relevant stakeholders. By 2004, less than half of the fisheries that had undergone pre-assessment decided to proceed to a full assessment (Bridgespan Group 2004, p. 4).

In a full assessment, the certifier appoints a team comprising experts in fishery-stock assessments, ecosystems and fishery management, who ascertain if applicant fisheries meet the MSC certification requirements. The detailed assessment used to involve the construction of an assessment tree for each applicant fishery. This rulemaking task was given to the expert assessment teams, granting them significant influence over the elaboration and interpretation of MSC's general principles and criteria. The assessment tree comprised performance indicators relating to the MSC principles and criteria, along with scoring guideposts defining the level of responses needed to achieve passing marks and reflecting the specific characteristics and concerns associated with the fishery applying for certification. Differences among assessment trees were intended to allow for variation in the conditions of each fishery. MSC observed, however, that this approach allowed expert assessment teams too much leeway in their interpretation of the principles and criteria; different teams developed different assessment trees for similar fisheries (MSC 2008a). Critics also observed significant variation in the assessment scores awarded by different teams across like fisheries seeking MSC endorsement (Ward 2008).

To address these problems, MSC introduced a new fisheries assessment methodology in July 2008. The new methodology was described by MSC as the 'biggest change' in the program since the standard was created in the late 1990s (MSC 2008a). At the heart of the new methodology is a default assessment tree that must be used in all future assessments (MSC 2008b). Rather than developing individual assessment trees for each fishery, expert assessment teams are now required to use the performance indicators and scoring guideposts of the default assessment tree as the basis for scoring all fisheries. Whereas the client for certification must provide the assessment team with the information that allows the team to score the fishery, team members are required to interview relevant stakeholders and to consider all concerns relating to the management and sustainability of the fishery. The team members then score the fishery according to the assessment tree and issue a preliminary report for peer review and public comment. By introducing a default assessment tree, MSC has sought a balance between the need for individual assessments of each applicant fishery and the need for consistency in the application of its principles and criteria. Yet, expert assessment teams still have significant discretion in scoring fisheries according to the assessment tree. It remains to be seen if the new fisheries assessment methodology will ensure greater consistency in the scoring of similar fisheries seeking MSC certification.

There is significant stakeholder involvement throughout the fishery assessment process. Any stakeholder can, in principle, provide input into the process, and team members must demonstrate that these comments have been considered in their final report. The assessment team will also arrange a number of meetings with various stakeholders throughout the process.[9] At the culmination of this process, the certifier decides if a fishery is to be certified. Stakeholders who have been involved in the assessment may object to the certifier's decision, in which case a complaints procedure is activated. The MSC certificate is valid for five years, and the fishery is subject to annual third-party audits of fishing operations. Before the end of the five-year period, the fishery must undergo another major assessment in order to renew its certificate.

A so-called chain-of-custody assessment must be conducted for the entire fish and fisheries product supply chain in parallel with or following the assessment of the fishery. The purpose of this assessment is to track the origin of the products through every stage of the supply chain, to assure end consumers that products carrying the MSC logo originate with a certified fishery. In order to use the logo on a product, the client undergoing certification must hold a licensing agreement with MSC. Because supply chains for seafood products are diverse and are typically lengthy and complex, chain-of-custody assessments can be challenging (May et al. 2003, p. 15). Provided that clients obtain a licensing agreement, they can use the logo on material other than a product containing seafood ('off-product') without having a chain of custody certificate, thus permitting companies such as restaurants and retailers to make general claims about their support for MSC (Highleyman et al. 2004, p. 8).

GOVERNMENT AND PRODUCER RESPONSES TO MSC

In spite of MSC's linkage to the FAO Code of Conduct and other fisheries agreements, and given the long history of international fisheries governance, certain European governments have been dubious about the scheme and have questioned the right of non-state bodies to govern common-pool resources such as fish stocks (Gulbrandsen 2006). Seeing MSC as an attempt to create a private transnational management regime beyond national jurisdiction, these governments argued that non-state actors had neither the necessary experience nor the mandate to govern fisheries. Unlike most standardization bodies, MSC allocates no preferred position to governments, which they treat like all other stakeholders – NGOs, fishers, producers and retailers, for instance.

In 1996, partly in response to the creation of MSC, the Nordic Council of Ministers formed a Nordic project group mandated to assess standards for sustainable fish production (Auld 2007). Based on its view that MSC was lacking credibility within the fisheries sector and among governments (Stokke 2004), the Nordic Council subsequently became a central proponent of an FAO-led labeling scheme (O'Riordan 1998). At the initiative of the Nordic countries, FAO's Committee of Fisheries discussed the practicality and feasibility of fishery certification and labeling at its biannual meetings in 1997 and 1999, and an FAO technical consultation examined the matter (FAO 1998). At neither meeting, nor through consultation, was agreement reached about the course of action that FAO should take (FAO 1999). Led by Mexico, the Latin American countries argued that eco-labeling in fisheries should be dealt with exclusively under the auspices of WTO rather than under FAO. Based on their experiences with US eco-labeling provisions to protect dolphins and sea turtles, developing countries were deeply skeptical of an FAO-led labeling scheme, which they believed would limit market access for their fisheries (O'Riordan 1998). Mexico, for example, had experienced a plunge in tuna exports to the

USA following the US regulations on dolphin-safe labeling. Agreement on nonbinding guidelines for labeling thus seemed more likely than did agreement on any labeling scheme with government involvement.

At another Committee of Fisheries consideration of the issue in 2003, the developing countries did not object to the development of *voluntary* guidelines for eco-labeling of fish products from wild-capture fisheries (Hoel 2006, p. 352). Experts and governments subsequently drafted a set of such labeling guidelines during a series of FAO expert and technical consultations. These nonbinding guidelines, issued by FAO in 2005, stated that labeling programs should include objective third-party fisheries assessments using scientific evidence; transparent processes with extensive stakeholder consultation and opportunities for complaints and rules for adjudication; and standards based on the sustainability of target species, ecosystems and management practices (FAO 2005).[10] Although the guidelines fell short of prescribing mandatory requirements for the use of eco-labels, they represented a step toward increased government influence over non-state labeling schemes. In essence, the creation of labeling guidelines was an effort by certain governments to regain control of an issue area dominated by non-governmental actors.

In March 2005, MSC issued a statement welcoming the FAO guidelines (MSC 2005). In order to comply fully with the guidelines, the program had to separate the standard-setting and accreditation functions. MSC thus outsourced accreditation decisions to Accreditation Services International – an independent organization created by FSC to accredit third-party certification bodies. Similar to FSC, MSC's approach to accreditation has thus evolved from an in-house process conducted by the MSC Accreditation Committee, to one controlled by a separate organization. Furthermore, it was obliged to make its procedure for handling objections to fishery assessments independent of the certification program. MSC reported that both these changes were implemented by September 2006 (MSC 2006b). In sum, the FAO guidelines seem to have consolidated MSC's position as the leading eco-labeling scheme for wild-capture fisheries, making it more difficult for potential competitors to create a scheme with equally strong requirements.[11] Nonetheless, some MSC competitors surfaced in response to the scheme.

As mentioned, fishers and fisheries industries in several countries were initially skeptical of MSC because of the central role of WWF and Unilever in creating the scheme. Swedish fishers rejected MSC certification and decided instead to partner with the Swedish organic labeling organization, KRAV, to develop standards for fisheries certification. Focusing primarily on organic standards, KRAV had no experience of developing standards for wild-capture fisheries, but it was regarded as a well-known labeling organization that could organize the standard-development process, while lending credibility to the label (Boström 2006b). The process began in 2000, when KRAV offered to coordinate the work that was being undertaken with standards, which until then had lacked effective coordination and leadership. From there, the work progressed quickly, with a final standard proposal for the eco-labeling of wild-capture fish and shellfish being presented in 2003 and approved by KRAV in 2004. Whereas the MSC principles and criteria relate only to fisheries activities up to but not beyond the point at which the fish is landed, the KRAV eco-label for wild-caught seafood also addresses the processing stage, including waste handling, packaging (only recyclable) and additives (only certain types permitted) (Thrane et al. 2009). A few Swedish shrimp and herring fisheries and two Norwegian fisheries (cod and haddock) that export to the Swedish market have been certified to the standard. Although several fisheries are in the assessment process, KRAV remains a regional scheme for the Swedish seafood market that does not challenge MSC's position as the world's leading fisheries certification scheme.

In the USA, an industry-led initiative was formed in 1997, when the National Fisheries Institute, a trade association of the US commercial fishing industry, launched its 'principles for responsible fisheries'. A year later the National Fisheries Institute created the Responsible Fisheries Society as a separate organization to facilitate the implementation of the principles (Auld 2007). Although the initiative was intended as an alternative to MSC (Carr and Scheiber 2002), it is primarily a guide to industry practices from extraction through to marketing rather than a third-party certification program. In short, MSC remains the only global certification program for wild-capture fisheries, but its decision to focus only on wild capture has enabled the formation and proliferation of a number of aquaculture initiatives (see Auld 2007; Lee 2008).

STAKEHOLDER INFLUENCE IN MSC

Certification programs like MSC and FSC are often portrayed as a new type of participatory, multi-stakeholder governance model that operates beyond the influence of governments. This governance model is praised by practitioners and researchers alike for its transparency, inclusiveness, broad representation of stakeholders and deliberative decision-making processes. Multi-stakeholder certification schemes are said to have a considerable democratizing potential through their ability to empower a range of stakeholders, including those who are marginalized in intergovernmental bodies and in traditional, state-driven rulemaking processes (for example Oosterveer 2005; Dingwerth 2007; Pattberg 2007). But does MSC's certification program and governance empower stakeholders who have traditionally had little influence on fisheries management and practices? Does the program contribute to more open and democratic fisheries governance?

The certification process offers stakeholders considerable leverage and provides a number of opportunities for deliberations and complaints. According to Rupert Howes, who became the chief executive of MSC in early 2005, the secretariat is doing its utmost to operate an inclusive, transparent and fully accountable program: 'I've spent 15–20 years in the non-government sector and never worked for an organization that devotes so much of its resources to governance and policy issues'.[12] A high level of transparency comes at a cost, however, in terms of finances and time spent on collecting and responding to stakeholder input. The requirements for inclusive certification processes could bury the organization in a bureaucracy of widespread consultation and cumbersome decision-making processes. From the viewpoint of the fishing industry, some environmental NGOs know only too well how to add time and costs to the assessment process and delay certification decisions. The complaint procedure represents another hurdle on the route to certification, and it has been invoked by NGOs in several cases. Brendan May, former chief executive of MSC, and his associates (May et al. 2003, p. 25) note that whereas MSC is arguably 'unique in its range of active stakeholders, its mandate, and the various levels at which input is sought', the organization is 'probably at a stage where any further significant increases in opportunities for input will jeopardize its aim of improving fishery management'.

Most stakeholders seem to agree that certification proceedings in MSC are inclusive and transparent, offering them a number of possibilities to engage in the process. Giving stakeholders access to information and a voice in the assessment process does not necessarily provide them with influence over assessment outcomes, however. Indeed, environmental and social stakeholders have complained that certifiers do not adequately address their concerns in fishery assessments and that they are too attentive to the needs of the fishing industry. The authority delegated by MSC to the certifiers arguably comes at the expense of the ability of social and environmental stakeholders to affect decisions. The assessment process is paid for by the applicant fishery, which, in the process, becomes the certifier's client. Because expert assessment teams appointed by the certifier score applicant fisheries and determine certification outcomes, they have considerable decision-making power within the program. Yet, apart from being accredited by an independent organization, certifiers – and the expert assessment teams they appoint – are not directly accountable to social and environmental stakeholders or to the general public. As mentioned, stakeholders may object to the certifier's decision, thus activating a complaints procedure, but as of December 2009, no complaints have resulted in the suspension of a certification. Despite the open and transparent certification proceedings in MSC, environmental and social stakeholders have not been able to successfully challenge certification of what they regarded as controversial and unsustainable fisheries. In this sense, we see that the high level of procedural transparency in MSC has not empowered stakeholders to affect certification outcomes or to hold certifiers to account (Auld and Gulbrandsen 2010).

On the broader issue of governing the program, MSC remains controversial because of similar concerns raised about power asymmetries in decision-making processes. Granting ultimate decision-making authority to the Board of Trustees rather than to the Stakeholder Council meant that social and environmental stakeholders have limited power to influence what MSC ought to be doing as a certification program. Power asymmetries among stakeholders within MSC are accentuated by variation in their access to funds and other power resources; whereas large-scale industrial fisheries have the ways and means to participate in decision-making processes and to influence outcomes, social and environmental stakeholders often lack the capacity to influence outcomes or even to send representatives to meetings. Contrary to the hopes of some of their founders, certification schemes risk preserving and even accentuating existing power asymmetries – between economic and non-economic stakeholders, between the resourceful and the marginalized, and between developed and developing countries – rather than empowering marginalized stakeholders. These issues are further discussed in Chapter 7.

CONCLUSIONS

This chapter has shown the ways in which MSC's origins, scope and approach to governance have shaped debates about fisheries certification and the subsequent development of the program. Whereas FSC involved direct membership and made the choice to devolve authority to national affiliates, MSC began with a more centralized approach to governance. Having learnt from FSC's experiences with an inclusive governance arrangement, WWF and Unilever decided to create a more streamlined approach to stakeholder involvement and governance. In order to prevent cumbersome and time-consuming decision-making processes, they chose to avoid granting decision-making power to a membership and to national affiliates. MSC's founders did include national affiliates for the purpose of outreach, but they did not devolve rulemaking authority to those affiliates. Rather, the task of making MSC's general principles and criteria locally appropriate was given to the certification bodies that were to assess individual applicant fisheries. Because certification bodies – and the expert assessment teams they appoint – also determine the outcome of the certification process, they have significant power within the program. These bodies are accredited by an independent organization, Accreditation Services International, but they are not accountable to stakeholders or to the general public.

Although WWF and Unilever included other stakeholders through workshops and consultations on appropriate standards, the streamlined approach to stakeholder involvement courted controversy from the start. As a result, the partners quickly moved to establish MSC as an independent organization. MSC subsequently established its Advisory Board, later replaced by the Stakeholder Council, to include a wide range of stakeholders in governing the program. These governance reforms did not, however, come with a similar power shift that would empower stakeholders in rulemaking processes within the scheme. Whereas the Stakeholder Council has an advisory role, ultimate decision-making authority remains with the Board of Trustees, which is self-recruiting and functions much like a corporate board of directors. In this respect, MSC differs from FSC's membership-based governance model, where the highest decision-making authority rests with the General Assembly. MSC also differs from FSC in its decision not to address social issues in its principles and criteria. As discussed in Chapter 7, this decision meant that in its early years of operation, MSC paid little attention to the access of developing country fisheries to the program, and, consequently, few of those fisheries have obtained MSC certification.

Giving certification bodies instead of national program affiliates the task of making MSC principles and criteria locally appropriate meant that since its inception, MSC has had a more centralized governance structure than FSC has had. Despite governance reforms and changes within the program, MSC has upheld this approach to ensure efficient and flexible decision-making processes within the organization. In this respect, we see that imitation of a specific governance model is likely to be mixed with innovation as a result of adaptation to a different context, power struggles over whose interest the model is to serve and variation in the configuration of interests in different policy fields or sectors. In the case of fisheries certification, MSC's governance structure, standards and certification requirements continue to raise questions about the extent to which MSC has empowered environmental and social groups – including marginalized stakeholders in developing countries – with a stake in fisheries governance. The next chapter examines this issue and other effects of fisheries certification.

NOTES

1. Full title: 'Agreement for the Implementation of the Provisions of the United Nations Convention on the Law of the Sea of 10 December 1982 relating to the Conservation and Management of Straddling Fish Stocks and Highly Migratory Fish Stocks'.
2. Full title: 'Agreement to Promote Compliance with International Conservation and Management Measures by Fishing Vessels on the High Seas'.
3. www.seafood.audubon.org/ (accessed October 21, 2008).
4. See: www.montereybayaquarium.org/cr/seafoodwatch.aspx.
5. Since 2008, the board has had two permanent seats each from the seafood industry, conservation NGOs and the market sector (food service, retailers, brand marketers).
6. In March 2008, the Board of Trustees decided that the Stakeholder Council should have a maximum membership of 40 (Summary of MSC Board of Trustees Meeting, March 12, 2008).
7. Interview with Rupert Howes, MSC Executive Director, May 23, 2006 (on file with author).
8. Interview with Alice McDonald and Daniel Suddaby, MSC Fishery Assessment Officers, May 23, 2006.
9. Interview with Alice McDonald and Daniel Suddaby, MSC Fishery Assessment Officers, May 23, 2006.
10. Full title: 'The FAO Guidelines for the Ecolabelling of Fish and Fisheries Products from Marine Capture Fisheries' (FAO 2005).
11. Interview with Rupert Howes, MSC Executive Director, May 23, 2006.
12. Interview with Rupert Howes, MSC Executive Director, May 23, 2006.

7. The adoption and impact of fisheries certification

In 2009, MSC celebrated its first ten years as an independent and operational certification program. Given the experience of a decade of operations, it is interesting to undertake a critical examination of MSC's achievements and effectiveness in delivering on its promises to reverse the decline of fish stocks and enhance marine conservation. This chapter examines the effectiveness of MSC by looking at producer adoption of the program, patterns of adoption, the effects of the certification process, and criticisms of and challenges for the program. As in Chapter 4, particular attention is paid to the question of how patterns of adoption influence the effectiveness of the program, and a distinction is made between the direct effects of a certification scheme and the broader consequences that flow from the emergence of that scheme. Using a narrow definition of effectiveness, fisheries certification would be judged effective if it contributed directly to the resolution of the problems it was created to address (overfishing, environmental harm resulting from fishing). Yet a broad conception of effectiveness would consider not only direct effects, but also environmental, social and economic effects that were not necessarily intended or anticipated. This chapter examines both narrow, problem-solving effectiveness and the broader consequences of fisheries certification.

The chapter begins with an examination of patterns of producer adoption of the certification program. The second section seeks to explain patterns of adoption, paying particular attention to the types of fisheries that participate in the program and the extent of participation in developing countries. In the third section, the environmental benefits and impacts of fisheries certification are assessed. The fourth section discusses criticisms of and challenges to the MSC certification program. The conclusion reflects upon the lessons learned from the MSC experience, arguing that we need more research on the intersection of private and public efforts to address overfishing and environmental harm from fishing.

PATTERNS OF ADOPTION

Calling MSC-labeled fish and fisheries products 'The Best Environmental Choice in Seafood', WWF and Unilever have actively promoted the label among leading retailers in key countries. When MSC was created in 1996, Unilever pledged that by 2005 it would be buying all the fish for its several brands, such as Iglo, Birds Eye and Findus, from sustainable sources (Weir 2000). Less than one year later, in April 1997, the UK supermarket chain Sainsbury's became the first food retailer to commit to sourcing MSC-certified fish products. The supermarket chains Tesco and Safeway soon followed Sainsbury's lead, rendering the three largest food retailers in the UK supporters of MSC (Constance and Bonanno 2000). MSC also launched a campaign in 1997 to provide a wide range of stakeholders with the opportunity to endorse MSC publicly, by signing a letter of support. Signatories included food retailers, fish buyers, fish processors, conservation organizations and a few fish-workers' organizations.

The founders of MSC hoped that whereas consumers may not pay more for eco-labeled fish products, they will prefer it in the marketplace over a product that lacks this type of environmental sustainability credentials. Offering consumers a real choice, however, requires participation from a critical mass of producers. Sufficient producer participation to penetrate markets was a significant challenge for MSC in its early years of operation. When the big supermarket chains in the UK announced their support for MSC, no fisheries had yet been certified, meaning that supply felt short of demand. Rupert Howes, MSC's executive director, explained how limited supply was a major challenge for the scheme at the time:

> In the early years, MSC suffered from no supply, no market; no market no supply. A lot of big retailers would say we're under huge pressure to demonstrate corporate social responsibility (. . .) they didn't want Greenpeace in their car park, saying that they were stocking unsustainable seafood. They were supportive of MSC, but they couldn't build a procurement strategy around the program.[1]

Fortunately for MSC, the situation improved quickly. A milestone was achieved in 2000 when, after 15 months of assessment, Western Australia's rock lobster fishery became the first to be certified to the MSC standards (MSC 2000a). This certification was closely followed by the Thames-Blackwater Herring Driftnet fishery in the UK, a small-scale enterprise of fishers supplying Sainsbury's Essex stores (Fowler and Heap 2000, p. 140). Later in the year, MSC certified the large-scale Alaska salmon fishery, comprising thousands of operators catching more than 350,000 tons of salmon annually (Hoel 2006, p. 362). Certification of the Alaska salmon fishery gave MSC a much-needed entry into the US market. Whole Foods Markets, a large retailer of organic foods, subsequently committed itself to sourcing MSC-certified seafood to its customers, as did Trader Joe's, Shaw's and Legal Seafoods (MSC 2000b). Less than a year after the certification of Alaska salmon, Whole Foods Markets became the first supermarket chain in the USA to offer MSC-certified salmon to end consumers (MSC 2001b).

MSC continued expanding in 2001, with the certification of two small-scale fisheries and the large-scale New Zealand hoki fishery, one of the country's largest fisheries and the most valuable fish export. Although critics alleged that some of the fisheries certified earlier, particularly Western Australia's rock lobster fishery, did not meet the MSC principles and criteria, (Sutton 2003), there had been no formal objections. In contrast, the hoki certification was immediately challenged by the Royal Forest and Bird Protection Society – New Zealand's leading environmental group. They alleged that there were numerous knowledge gaps regarding the impact of the hoki fishery on ecosystem structure and function. Specific concerns included the risk of overfishing under existing harvest levels, lack of an ecological risk assessment, significant by-catch of seals by deep-water hoki trawling and lack of compliance with the New Zealand Fisheries Act (Highleyman et al. 2004). An appeals panel appointed by MSC noted that there were ten corrective action requirements included in the certification of the fishery; the panel added several recommendations for improving the management situation, but decided to uphold the certification. Although this assessment did little to calm the major environmental groups in New Zealand, who continued to protest the decision, the hoki fishery has kept its certificate.[2] As a result of the hoki controversy, however, in 2002 MSC established a complaints procedure to better handle conflicts over certification decisions (Hoel 2006).

A major boost in supplies of certified volumes came in 2004, after a nearly four-year assessment process culminating in the highly controversial certification of Alaska pollock – by far the largest certified fishery in the world. MSC certified two stocks – Bering Sea and Gulf of Alaska pollock – both of which provoked complaints. In the case of Bering Sea pollock, for example, four federal court decisions since 1998 have found the fishery out of compliance with federal law that requires the National Marine Fisheries Service to consider the fishery's impact on other ecosystem components or endangered species (Highleyman et al. 2004). Nonetheless, MSC eventually decided to uphold the certification of Bering Sea and Gulf of Alaska pollock.

Three other fisheries were certified in 2004: South Georgia Patagonian toothfish (a fish species that is marketed and sold as Chilean sea bass in the USA and Canada), Mexican Baja California red lobster (the first community-based fishery to be certified) and the large-scale South African hake trawl fishery. One certification – for Patagonian toothfish around the island of South Georgia in the southern Atlantic Ocean – resulted in an official appeal from Greenpeace, the Sierra Club, the Natural Resources Defense Council, the National Environmental Trust and the Antarctica Project. They jointly filed a complaint, asserting that certification of South Georgia Patagonian toothfish would lend false credibility to all Patagonian toothfish fisheries and exacerbate the problem of illegal fishing (Highleyman et al. 2004; see also Agnew 2008), which is estimated to constitute more than 50 percent of the total catch of Patagonian toothfish (Lack and Sant 2001). Another concern was that the certification of South Georgia Patagonian toothfish could disrupt the consensus-based decision-making process in the Commission for the Conservation of Antarctic Marine Living Resources – a multilateral organization tasked with overseeing fisheries management systems for all Southern Ocean waters (Highleyman et al. 2004). After reviewing the complaint, however, the objection panel decided to uphold the certification. Thus far, no complaint in MSC has resulted in a withdrawal of a certificate that has been awarded to a fishery.

By late 2004, the total catch size of the fisheries engaged in the program was about 1.8 million tons or nearly 4 percent of the world's edible capture fisheries (Hoel 2006, p. 362). One fishery – Alaska pollock – accounted for approximately two-thirds of the volume of MSC-certified fish. As a result of this certification, the proportion of Unilever's European fish products made from MSC-certified fish jumped from 4 percent in 2004 to 46 percent in 2005 (Unilever 2005). Furthermore, MSC could rightfully claim to be a major player in the global whitefish market (hoki, hake and pollock), supplying about one-fifth of the traded volume (Hoel 2006, p. 365). Nonetheless, because of an insufficient supply, Unilever did not attain its goal of sourcing all fish for its frozen fish brands from MSC-certified sources by 2005.

With a critical mass of certified fisheries on board, the program was ready for further expansion in the consumer market. A major breakthrough came in January 2006, when Walmart, the world's biggest retailer, announced a commitment to source all its fresh and frozen seafood supplies in North America from MSC sources within five years (MSC 2006c). Following on the heels of this announcement, ASDA, Walmart's UK subsidiary, declared that it would sell only MSC-certified fish in all its stores within three to five years (MSC 2006d). According to

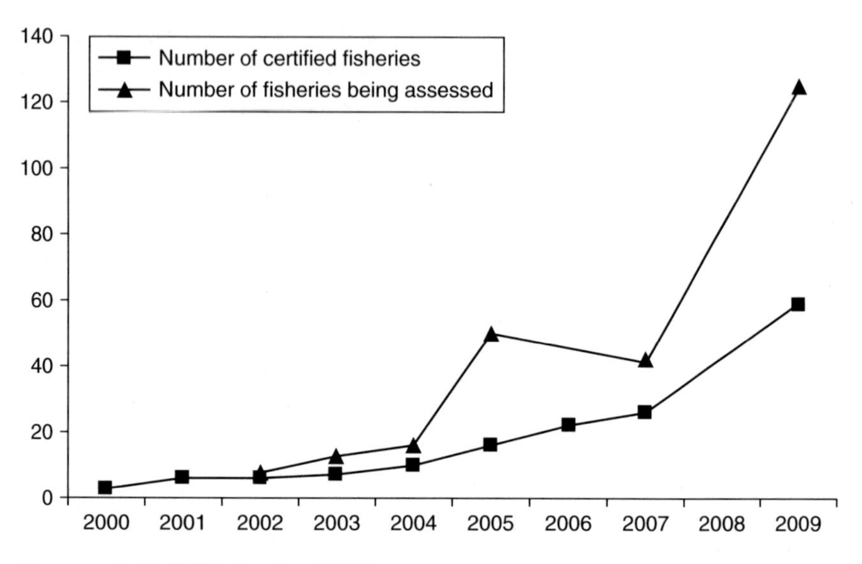

Sources: Auld (2009); data for 2009 from www.msc.org.

Figure 7.1 Number of certified fisheries, and fisheries in assessment, by year

Rupert Howes, the Walmart commitment catalyzed other retailers to look at their own commitments to MSC:

> I think there is a trend now for retailers in both Europe and North America to build MSC more visibly and strategically into their seafood procurement policies. Some retailers are becoming far more proactive, wanting to be the first to launch a new MSC species and product to the market. Others want to go down their own supply chains to encourage the fisheries they ultimately source from to move forward into the independent and scientific assessment process.[3]

By the end of 2009, 59 fisheries were certified to the MSC standard and another 125 were in the assessment stage, accounting for more than 7 percent of all wild-caught seafood sales. Together the MSC-certified fisheries record annual catches of more than 5 million tons of seafood, representing more than 42 percent of the world's wild salmon catch, 40 percent of the world's edible whitefish catch and 18 percent of the world's lobster catch. It is essential, however, to consider patterns of adoption in assessments of the environmental impact of certification programs. Such patterns are examined in the next section. Figure 7.1 shows the number of certified fisheries and fisheries in assessment from the first certifications in 2000 to December 2009.

EXPLAINING PATTERNS OF ADOPTION

Unlike the case of forestry, certification pressures in the fisheries sector have generally not originated from NGO targeting of producers. Rather, such pressures have come from NGO alliances with major buyers – the WWF-Unilever partnership in particular – and their strategic work to create a market for sustainable seafood. Large companies have, to some extent, dictated the terms of buying and selling arrangements up and down the supply chain, resulting in demand for certification. Unilever's initial pledge, announced in 1996, to buy all fish for its brands from sustainable sources, clearly convinced other buyers, as well as suppliers, to become involved in MSC. With backing from WWF and Unilever, MSC itself has played a vital role in harnessing market forces to encourage fisheries to certify. Support from the UK supermarket chains, Sainsbury's, Safeway and Tesco, was forthcoming largely because of collaboration with MSC and its two founders. Similarly, Walmart's 2006 commitment to source MSC-labeled products a decade after the inception of the program, was largely a result of MSC's work to build a larger market for sustainable seafood. The Walmart commitment pressured its supplier to become involved in MSC, and could potentially transform the seafood industry, not only by increasing demand for MSC but also by forcing other retailers to rethink their purchasing policies. Of course, major retailers are also motivated by the desire to avoid naming and shaming by environmental NGOs; the initial Greenpeace campaign against Unilever was successful in that it moved the company to collaborate with WWF to create MSC (Conroy 2007, p. 219). Yet, compared to forest certification, there has been relatively little advocacy group targeting of companies to pressure them to certify or to source only MSC-labeled seafood.

The commitment of major buyers to source only MSC-labeled seafood could create market advantages for fisheries in developed countries and disadvantages for fisheries in developing countries. Recall from Chapter 2 that because participation in certification schemes is voluntary, it is possible that only those producers facing minor compliance costs will opt in. If producers who face substantial compliance costs were to opt out of certification schemes, the net effects of certification initiatives would be low. This selection problem is evident in fisheries certification; those fisheries that currently meet the MSC criteria share several key characteristics and differ from the majority of the world's fisheries. Two types of fisheries dominate the scheme: small-scale and large-scale fisheries in developed countries. There are few certified intermediate-sized fisheries (Kaiser and Edward-Jones 2006). As observed by Hoel (2006, p. 354), current certification requirements may favor fisheries in industrial coastal states because they can afford the certification costs and have the means to participate in the assessment process. In addition to uncertainty about the market benefits accruing from certification, fisheries considering whether or not to engage in a pre-assessment often perceive the cost of the certification process as a major obstacle.[4] The full certification process can be time consuming, costly and demanding for the fishery undergoing assessment, as seen in the four-year assessment of Alaska pollock (Gulbrandsen 2005b). The Alaska pollock experience is not the norm, however; more commonly, the assessment process lasts about 12 months.[5] In order to comply with the standard, fisheries must undergo assessment and logo licensing costs. In addition, they must often implement a number of costly changes in their operation (changing gear, reducing by-catches of nontargeted species and disbanding fishing units, for example), which may far exceed the short-term costs of the assessment process (Hoel 2006).

Kaiser and Edward-Jones (2006) examined the key features of the first 11 MSC-certified fisheries. They found them to be highly selective of their target species; to have stocks that occur within known areas, for which there are exclusive national access rights; to tend to have limited access; to be well-regulated and enforced; and often to be co-managed by governments, scientists and fishers (Kaiser and Edward-Jones 2006, p. 394). In contrast, most fishers in most regions of the world have no significant input into the management process; they share the fish resources with unassociated fishers or with multiple fishers from other nations, and have little control over the setting of fishing quotas (Kaplan and McCay 2004). Many fishers, in fact, are excluded from even considering MSC certification because of the actions of others that are beyond their control (Kaiser and Edward-Jones 2006, p. 394). With open-access resources, fisheries that meet most of the MSC criteria but share the fish resources with other fisheries that do not fish sustainably are effectively excluded from access to the label. One solution to this problem could be the formation of more fishing cooperatives, to enable collective action and co-management of the fish resources (Gelcich et al. 2005). The formation of fishing cooperatives, in turn, seems to require some form of government intervention to force fishers to work collectively and assume management responsibility for defined areas of the sea (Gelcich et al. 2005; Kaiser and Edward-Jones 2006). Given the nature of the fish resources, government intervention is also necessary to enable certification of a number of fisheries that currently fall short of the MSC criteria. In essence, because most fisheries are under the control of government bodies, fish stocks require government intervention for their conservation. One option for fisheries stakeholders, then, is to work with government regulators to change regulatory frameworks in ways that would allow certification of fisheries that do meet the MSC criteria (Leadbitter et al. 2006). Similarly, if governments believe that certification is vital for the economic viability and market access of the fishing industry, they may take the initiative to change management rules to allow for the certification of fisheries (Hoel 2006, p. 349).

Whereas the involvement of large-scale fisheries in MSC is largely market driven, community-based, small-scale and artisanal fisheries have a number of reasons for entering the certification process (Ward and Phillips 2008, p. 420). Demonstrating to government regulators that their fishery is well-managed through certification may ensure that the fishery is treated favorably in the allocation of catch quotas or in any other resource allocation decisions. In a study of the South African hake fishery, for example, Ponte (2008) found that MSC certification was used as a tool to prevent a redistribution of quotas away from the large (white-owned) deep-sea trawling sector to the smaller (black-owned) longlining sector. On the other hand, MSC certification has, in a few cases, empowered smaller producers and producers in poorer countries. Certification of the Baja California lobster fishery, the first fishery in Mexico and Latin America to be approved by MSC, has resulted in political empowerment of the fisheries cooperatives that were the clients for the assessment (Agnew et al. 2006). Enhanced bargaining power in negotiations with both the fishing and environmental authorities in Mexico has translated into the fulfillment of demands for electricity services for the fishing communities, the beginning of a rural road improvements program and other infrastructure services provided by the government (Phillips et al. 2008).

Except for the Mexican Baja California lobster fishery and a few other community-based fishery certifications, MSC has met with little success in developing countries. Indeed, a recurring line of criticism against MSC is the failure to certify small-scale, community-based and local fisheries in the South. As of mid-2009, only a few fisheries in developing countries have been certified, and the adoption of MSC-labeled products is largely limited to Europe, North America and Japan. These patterns of adoption have caused concern that labeling may restrict market access of non-labeled products from developing countries, with potentially severe consequences for their producers.

Sometimes portrayed as another instance of 'green protectionism', many developing countries see eco-labeling as a de facto barrier to trade, and have voiced their concerns in WTO deliberations such as those of the Committee on Trade and Environment (Gulbrandsen 2005b) and in other international organizations like the UN Conference on Trade and Development (UNCTAD 2007; Auld et al. 2008).

There are substantial barriers to achieving MSC certification in developing countries, ranging from lack of information and shortcomings of scientific data to financial costs. As early as 1996, questions had been raised in *Samudra* about whether or not certification served the interests of fishworkers in developing countries (*Samudra* 1996; Kurien 1996). Yet, in its early years of operation, MSC paid little attention to their needs (Ponte 2008). In this respect, we see how the decision not to address social issues in the MSC principles and criteria had negative consequences for the access of small-scale fisheries to the program. Grappling with the social side of fisheries management is clearly a key challenge to MSC, not least because concerns raised about social issues might have a negative impact on its credibility in the marketplace. Only more recently has MSC paid significant attention to social issues, as seen in the launch of its Developing World Fisheries Program. As part of this program, MSC is piloting a project to enable small-scale and data-deficient fisheries better access to its label.[6] The pilot project has developed guidelines to assist certification bodies involved in assessing such fisheries. These guidelines are intended to help certifiers use the type of information that may be available to small-scale and data-deficient fisheries, including traditional ecological knowledge and traditional management systems. Seven fisheries have used the guidelines in trial assessments to allow for practical testing, review and evaluation of their effectiveness. Since 2005, MSC has also worked with fisheries experts to develop a risk-based approach to the assessment of data-deficient fisheries, intended to complement its existing methodology for fisheries assessment. As a result of this work and the field trials, the program has developed what it now refers to as the MSC Risk-based Framework for fisheries assessment. A risk-based approach essentially aims at setting a lower threshold for the type and amount of information needed to certify small-scale and data-deficient fisheries, while maintaining stakeholder confidence that the fishing activity complies with the MSC principles and criteria. Apart from this project, the MSC Developing World Fisheries Program focuses on outreach and promotion, and has undertaken increasing involvement from developing country stakeholders in the program.

Notwithstanding the efforts to enable small-scale, community-based and local fisheries better access to the program, the costs of certification represent a major hurdle for these types of fisheries. MSC has worked with the Packard Foundation and with Resources Legacy Fund, a US-based NGO, to establish the Sustainable Fisheries Fund to provide grants or loans to fisheries organizations interested in certification.[7] Several fisheries have been certified with financial support from this Fund since its launch in 2002, but the grants are small and can only support such activities as ensuring stakeholder input into the fishery assessment process. The fund is not able to pay for full assessment costs or support other activities that might typically have received funding from development agencies

(Humphreys 2002, p. 25). Considering that most of the seafood in developing countries is consumed locally, in markets with little or no interest in eco-labeling, fisheries certification probably has limited potential to spread among the fisheries in these countries (Gulbrandsen 2006). The Asian seafood markets, by far the world's largest, have yet to see any breakthrough in seafood labeling. Given the current patterns of standards adoption and market adoption, fisheries certification would be more likely to modify the behavior of fishers in developed than in developing countries. In sum, current MSC certification requirements seem to favor two types of fisheries: small-scale fisheries in developed countries that are relatively easy to certify because they tend to have limited access and are highly selective of their target species, and large-scale fisheries that are well-regulated and can afford the comprehensive assessment process.

THE ENVIRONMENTAL IMPACTS OF FISHERIES CERTIFICATION

The *raison d'être* for the establishment of MSC was to reverse the decline of fish stocks and contribute to improvements in marine conservation worldwide, through a system of certification and labeling.[8] In the long run, continued support from consumers, philanthropic foundations and environmental organizations depends on the program's ability to deliver on its promises. Some observers argue that MSC's approach is promising in that it should help generate more sustainable consumption patterns (for example, Oosterveer 2005) and more sustainable fisheries management (for example, Leadbitter et al. 2006). Others are skeptical about its environmental impact, arguing that there is little evidence to suggest that it has arrested overfishing or delivered other ecological benefits (for example Jacquet and Pauly 2007; Ward 2008).

The ability of certification programs to modify fisheries practices and create better environmental outcomes ultimately depends on the assessment and certification processes. Analyses of environmental achievements by MSC certifications have yielded mixed results. A study commissioned by MSC to investigate the environmental gains resulting from its certification program reveals some of the effects of the assessment process (Agnew et al. 2006). The researchers examined a total of 62 certification conditions in the ten fisheries that, by late 2005, had been the subject of at least one post-certification audit, to determine if they would ultimately lead to environmental improvements. They then identified environmental gains indices for each certification condition and categorized the gains in five levels, ranging from no gain to the most desirable gains (called 'operational result' gains). Finally, they considered whether or not the environmental gains were caused primarily by the certification conditions, if they were ongoing anyway, or if they were a combination of the two.

The researchers found that all the certified fisheries have shown some environmental gain resulting from the certification process (Agnew et al. 2006). Yet, there was only one major ecological improvement related to the MSC certification process which was achieved in preparation for the assessment process: a reduction in endangered seabird by-catch in the South Georgia Patagonian toothfish fishery (Ward 2008). Reduced sea lion by-catch was identified in Western Australia's rock lobster fishery, but it was not directly related to the certification of the fishery. Although MSC certification did contribute to a reduction in fur seal by-catch in the New Zealand hoki fishery, it proved to be a temporary improvement (Ward 2008).

The researchers also reported some lessons learned, two of which are of particular interest (Ponte 2008). First, they identified the largest gains in areas that carried conditions for certification. Second, they argued that certification of difficult fisheries could be encouraged in order to maximize the environmental gains from the assessment process. These findings point to a dilemma for MSC (Ponte 2008), and indeed for all market-based certification programs (Auld et al. 2008). On the one hand, the stricter the certification requirements, the higher the potential environmental benefits from certification. On the other hand, there could be an inverse relationship between standard stringency and the adoption of a scheme by fisheries – particularly the difficult ones. Willingness to adopt a voluntary certification scheme with stringent standards is likely to be low among difficult fisheries, where the need for changing management practices is most urgent. Accordingly, there could be a tradeoff between environmental gain and producer adoption of a certification program (Ponte 2008; see also Auld et al. 2008; Raynolds and Murray 2007).

Looking further into the environmental benefits of certification, Ward (2008) investigated the distribution of all scores in the first 22 certified fisheries for each MSC principle. He found that one of the two main MSC certifiers systematically awarded higher scores for Principle 2 than did the other main certifier, indicating that 'the poorly expressed Principle 2 criteria are interpreted differently by these two certifiers, and applied differently in the various fisheries' (Ward 2008, p. 174).[9] Ward concluded that the MSC certification program has been unable to demonstrate major achievements in marine biodiversity conservation, reaffirming earlier contentions that the program has failed to contribute significantly to resolving environmental problems in the fisheries sector. Jacquet and Pauly (2007, p. 310), for example, claim that '[t]he MSC may create an incentive for industry to foster effective stock management, but has so far failed to demonstrably arrest the decline of fish stocks'. Others (for example Kaiser and Edward-Jones 2006; Ponte 2008; Ward and Phillips 2008) have reached similar conclusions.

In addressing the ability of current seafood labeling programs to achieve better environmental outcomes, Ward (2008) questioned how vested business interests between certifiers and their clients could result in flexible interpretations of the principles and criteria. Vested business interest in successful certification outcomes is a well-known challenge for credible forestry auditing (Ghazoul 2001; Gulbrandsen 2004). The competition among certifiers to secure assessment contracts may favor certifiers that are client-friendly in their assessments, thus lowering the bar for passing the assessments. Similarly, the cost of certification may create an incentive to use certifiers that can provide relatively cheap assessments. Although the flexible interpretation of principles and criteria was explicitly accepted by the MSC program, which until recently required certifiers to develop indicators and benchmarks for each fishery under assessment, the variation in assessment outcomes among certifiers could, in the long run, undermine trust in fisheries certification. As mentioned, in Chapter 6, MSC introduced a new fisheries assessment methodology to address this problem in 2008.

Forest auditing is relatively straightforward, because auditors can usually observe the direct effects of forestry operations in on-the-ground inspections. The nature of fish resources, on the other hand, renders them more of a challenge to credible auditing. There are often multiple access rights to shared fish resources, and many fish stocks are straddling and highly migratory. The absence of easily observable effects of noncompliance and the nonselective nature of many fishery harvest techniques further complicate the assessment process (Kaiser and Edward-Jones 2006). Characteristics of fish resource and fish governance make it difficult, therefore, to set standards that would lead to environmental improvements (Ward 2008). But as Leadbitter and Ward (2007) have discussed, it is possible to enhance the robustness of fisheries assessment systems, thereby avoiding lax assessment processes. As in the forest sector, stringent and comprehensive assessment criteria are likely to facilitate credible auditing, whereas lax or unclear criteria are likely to have the opposite effect.

Fisheries certification may also have consequences that were not necessarily intended or anticipated. Sutton (2003) describes how fishery managers in Western Australia's rock lobster fishery used the achievement of MSC certification to prevent the introduction of marine reserves in Western Australian waters, rejecting the need for fishing sanctuaries on the grounds that the fishery is certified. Another unintended consequence of certification is the favoring of fisheries in developed countries at the expense of fisheries and fishers in developing countries, where the costs of preparing for, paying for and participating in comprehensive certification assessments are often unaffordable. In addition, because many developing countries lack reliable scientific data on the state of their fisheries, they are excluded from even considering certification. As discussed, MSC has recognized this problem and is piloting the use of guidelines for the assessment of small-scale and data-deficient fisheries – the scheme's current keystone initiative in the developing world (Howes 2008, p. 100). Even so, the significant underrepresentation of developing country fisheries in MSC could challenge the credibility of the scheme, again highlighting the need to develop measures that would increase the participation of fisheries in developing countries.

In conclusion, there is little evidence to suggest that MSC has created significant environmental improvements. Ward and Phillips (2008, p. 433) argue that in their present development phase, seafood certification programs like MSC focus on uptake by resellers and promotion, whereas those programs tend to ignore the need for demonstrating environmental benefits of certification. It is not surprising, perhaps, that in its first decade of operations MSC has focused on producer adoption and market expansion. In the next decade, however, MSC must demonstrate the direct environmental impacts of its certification program. Unless MSC takes the challenge of demonstrating positive environmental impacts seriously, it risks defaulting to a marketing scheme for the seafood industry rather than serving as an institution for environmental change (Ward and Phillips 2008). In that event, it could still have a role as a program for enhancement of consumer awareness, much like seafood awareness campaigns, but it would not deliver on its promise to create environmental change through the certification of fisheries.

CRITICISMS AND CHALLENGES

Since becoming independent from its two founders, WWF and Unilever, MSC has had to source its own funding from a range of charitable trusts and foundations. Funding from the Packard Foundation played a major part in transforming MSC from an NGO-business partnership to a complex, multi-stakeholder governance scheme. Whereas some eco-labeling programs, such as the Nordic Swan (a successful eco-label in the Nordic countries) and the European Flower, are funded by governments, non-governmental schemes like MSC remain dependent on voluntary donations from private actors. According to the latest annual report, 77 percent of MSC's income in 2007–8 was charitable grants from trusts and foundations; financial support from government agencies, companies and individuals comprised 9 percent; and investments represented another 2 percent (MSC 2008c). Despite the goal of increasing the share of funding from logo licensing, only 12 percent of the income was logo license revenue. Most of the charitable grants came from trust and philanthropic foundations in the USA and the UK. In addition, MSC's founders, WWF and Unilever, continued to support the scheme financially, along with a few other NGOs and corporations. Two government agencies provided financial support for the scheme: the UK Department for Environment, Food and Rural Affairs and the Swedish International Development Agency (MSC 2008c).

When MSC began operations, there was reportedly 'a bit of naïve optimism that fisheries would be falling over themselves to be assessed, that they'd all be assessed very quickly, and the license revenue would sort of sustain the MSC'.[10] The idea was that logo licensing fees would generate the revenue to fund the operations (May et al. 2003, p. 30), but MSC is still extremely dependent on the support of a handful of organizations, with one source of funding – the Packard Foundation – contributing as much as 50 percent of the budget in some years (Hoel 2006, p. 360). This financial dependency obviously renders it vulnerable to changes in the objectives and priorities of its funders. MSC faces the challenge of avoiding the 'funder fatigue' sometimes experienced in organizations dependent upon private donors; many charities tend to jump onto new projects rather than supporting existing, well-established initiatives over the long run.

In spite of its accomplishments, MSC's practices and performance have continued to be intensively debated. The most severe criticism of the program came in 2004, when its principal funders commissioned two independent consultancy reports to examine certification practices and market acceptance. These reports were commissioned as a result of growing concern over MSC's certification practices in the environmental community. The certification of New Zealand hoki and the South Georgia Patagonian toothfish reportedly led MSC's funders to examine their involvement in the program and led to discussions at the highest level within MSC (Potts and Haward 2007, pp. 101–102). Many environmental organizations were also critical of MSC's practice of certifying fisheries before actual improvements in the fisheries occurred, as long as they committed to improving practices (Pearce 2003). They maintained that no fishery should be certified before it complied with all principles and criteria. On the issue of governance, environmental stakeholders believed that they were not integrated in the MSC decision-making structure in any meaningful way.

One of the reports, commissioned by three US-based marine conservation funders – the Oak Foundation, the Homeland Foundation and the Pew Charitable Trust – looked at fisheries certification and assessment practices, paying particular attention to Alaska salmon, New Zealand hoki, South Georgia Patagonian toothfish and Bering Sea pollock. The report identified a number of problems (Highleyman et al. 2004):

- MSC's claim of certifying sustainable fisheries was, in most cases, unjustified under the definition established by its principle and criteria.
- Principle 2, requiring fishing operations to maintain the structure, productivity, function and diversity of the ecosystem on which the fishery depends, was not met in many cases.
- Fisheries that were noncompliant with national fisheries regulations had been certified.
- Certifiers had 'too much flexibility' in applying the principles and criteria and in determining compliance thresholds, which resulted in inconsistencies and low thresholds.
- Key environmental stakeholders did not perceive MSC as credible because they believed that the scheme failed to 'include them in a substantive way'.
- Distrust from the environmental community increased with familiarity with MSC as an organization or through a specific certification.

The second consultancy report was commissioned by the Packard Foundation – by far MSC's biggest funder. This report was only slightly less critical, focusing on the lack of market acceptance and policy successes. It concluded that, as of the end of 2003, MSC had had limited market success: none of the major supermarket chains carried MSC-labeled products. The report noted that uptake of MSC-labeled products was largely limited to niche markets in Europe. The report also concluded that MSC had had few policy successes, noting that policy changes in fisheries management regulations linked to the scheme had occurred only in Australia and New Zealand (Bridgespan Group 2004).

We have seen that MSC has experienced considerable growth since the release of these two reports, indicating that criticism over lack of market acceptance was taken seriously. Indeed, the Board of Trustees and the secretariat took a series of measures to implement the recommendations from the reports. One such measure was the appointment of a new executive director, Rupert Howes, who committed to rebuilding a program that had almost lost support from key stakeholders in 2003 and 2004. By re-establishing contacts with critical environmental NGOs, raising funds for

MSC and working tirelessly to strengthen the organization, Rupert Howes has reinvigorated the program (Conroy 2007, pp. 217–18). Moreover, partly because of criticism from the funders, the Board of Trustees has evolved from an entirely self-selected committee to one that includes the chairs of the Stakeholder Council and the Technical Advisory Board. All but three members of the board are still self-appointed, however. As discussed earlier, MSC has also taken a number of steps to improve operations, including a review of its certification procedures, the introduction of a new fisheries assessment methodology to increase consistency across evaluations of similar fisheries and the development of guidelines for the assessment of small-scale and data-deficient fisheries. Yet, the impact of the scheme on fishing practices and ecosystems affected by fisheries remains contested.

It is evident that MSC depends crucially on continued goodwill from supportive trusts and foundations. Work on the role of US philanthropic foundations in the expansion of FSC has shown that they have been influential not only as sources of funds for the program, but also as critical supporters and coordinators of NGO campaigns pushing producers to certify (Bartley 2007). In the case of MSC, philanthropic foundations have primarily been important as funders of the program rather than as backers of NGO efforts to convince fisheries organizations to certify – possibly because, unlike FSC in the forest sector, MSC is free of direct competition from other certification schemes. But if MSC is unable to demonstrate significant environmental benefits from certification, it runs the risk of losing support from environmental NGOs and from the foundations that finance the large share of its operations. The environmental organizations and foundations that currently back MSC must see evidence that certification makes a difference – not only in the marketplace, but also in environmental problem solving in the fisheries sector.

CONCLUSIONS

A number of process improvements in MSC-certified fisheries indicates that certification can benefit fisheries management and practices. When we turn to the issue of environmental impact, the link to certification programs becomes more tenuous. Certification alone is unlikely to resolve the dire problems of overfishing and depleted fish stocks; government-sanctioned marine reserves, rules restricting access to fish resources, stringent distributive schemes and the curtailment of IUU fishing must all be part of the solution. The regional and global scale of overfishing and depleted fish stocks is a significant challenge to certification as a tool for addressing problems that are rarely contained within a single fishery. Moreover, patterns of adoption continue to raise questions about effectiveness. Being highly selective of their target species, well-regulated and enforced, and with limited access rights, certified fisheries differ from the majority of the world's fisheries. Fisheries in developing countries are underrepresented in the program. Having paid little attention to this problem in its early years of operation, MSC has recognized that certification of small-scale and data-deficient fisheries in development countries represents a major challenge. The scheme has developed a risk-based approach to certifying such fisheries, as part of a special program seeking to increase adoption of the scheme in developing countries. Yet patterns of producer adoption of the program continue to raise questions about its legitimacy and effectiveness, leading to the worrying possibility that certification can indeed marginalize smaller fisheries and fisheries in developing countries. Large-scale fisheries in developed countries and other powerful economic actors have so far had much greater access to MSC than have small-scale, community-based and local fisheries in developing countries.

Although MSC has been operational for more than a decade, it may still be too early to identify the environmental impact of certification on marine ecosystems and oceans. At this early stage of fisheries certification, perhaps the least tentative conclusion should be that MSC has succeeded in achieving considerable producer and market adoption during its first decade of operations. With respect to the environmental effects of fisheries certification, a critical area of study is the intersection of private and public efforts to resolve the problem of overfishing and decrease the environmental harm resulting from fishing. Voluntary certification programs in the fisheries sector will essentially remain as supportive tools to intergovernmental and government regulations. Because it will continue to be the role of governments to agree upon and enforce fishery management regimes, certification programs must engage more directly with existing government regulations and policies. More research is needed, therefore, to address the question of how voluntary certification and government regulations can interact and be mutually supportive in reversing the decline of fish stocks and creating improvements in marine conservation worldwide.

NOTES

1. Interview with Rupert Howes, MSC Executive Director, May 23, 2006.
2. The New Zealand Hoki fishery certificate was renewed in 2007 (MSC 2007).
3. Interview with Rupert Howes, MSC Executive Director, May 23, 2006.

4. Interview with Rupert Howes, MSC Executive Director, May 23, 2006.
5. Interview with Alice McDonald and Daniel Suddaby, MSC Fishery Assessment Officers, May 23, 2006.
6. The full project name is Guidance for the Assessment of Small-scale and Data-deficient Fisheries (GASSDD).
7. See www.resourceslegacyfund.org/pages/p_fish.html.
8. See www.msc.org/.
9. The criteria in Principle 2 relating to conservation issues have been criticized for their broad and highly aspirational terms, which render them unlikely to be achievable by any wild-capture fishery (Sutton 2003; Ward 2008).
10. Interview with Rupert Howes, MSC Executive Director, May 23, 2006.

8. The spread and institutionalization of certification programs

Since the formation of FSC in the early 1990s, the certification model has spread to a number of sectors and industries. Some certification initiatives mimicked established programs like FSC, whereas other initiatives began as separate processes. Yet, all are strikingly similar in their organizational design and governance processes and procedures. In every case, the emergence of the certification program was part of a broader shift from government command-and-control regulations toward voluntary approaches to environmental problems. The lack of effective multilateral and domestic regulations addressing such transnational problems as forest degradation, fisheries depletion and sweatshop labor practices made environmental and social groups turn to the business sector itself. NGO-backed certification programs sought to achieve legitimate rule-making authority through multi-stakeholder governance arrangements; yet, as we have seen in some sectors, such as forest, producer-backed certification schemes emerged in response to NGO efforts to regulate producers and industries.

This chapter examines the ways in which certification programs have evolved and spread across sectors. The first section examines the proliferation of non-state certification schemes, with particular attention to the policy entrepreneurs who have carried facets of the certification model from the forest sector to several other sectors. The second section examines the formation of alliances among certification schemes and between certification schemes and international organizations, in an effort to achieve legitimacy as standard-setting organizations. Focusing on FSC, the third section turns its attention to the challenges of achieving balanced stakeholder participation over time in the governance of multi-stakeholder certification programs. Drawing on a study of the Fair Trade movement, the last section in this chapter outlines key challenges for social and environmental certification programs.

SPILLOVERS TO OTHER SECTORS

FSC's founders turned to the International Federation of Organic Agriculture Movements (IFOAM) and the International Union for Conservation of Nature (IUCN) as organizational models for the FSC chamber system (Elliott 1999). FSC has, in turn, become an organizational model for other certification programs. We have seen that FSC's success served as a major motivation for WWF to establish a similar certification program in the fisheries sector. As discussed in Chapter 6, the creation of MSC was not simply a case of copying FSC's organizational model. Having learned from FSC's experiences, the founders of MSC mimicked some of its features, while strategically avoiding other features. This selective mimicry resulted in a distinct organizational model, although WWF was careful to draw attention to the similarities between the two schemes in the hope that FSC would lend credibility to MSC (Auld et al. 2007).

In addition to the spillover from forests to fisheries, the certification model has spread to several other industries and sectors, including sustainable tourism, the marine aquarium trade, palm oil production, soy production and parks management (Honey 2002; Conroy 2007; Auld et al. 2007). Some certification initiatives had largely independent roots; labor standards and forestry standards emerged roughly at the same time, for example, but those working on the respective schemes had little knowledge of what was happening in the other sector (Bartley 2003). In other cases, a few policy entrepreneurs played a critical role in spreading the certification idea across sectors and industries. As outlined by Auld et al. (2007), three entrepreneurial groups served as organizational 'carriers' of the certification model, helping to transport it to multiple sectors: (1) environmental NGOs, (2) professional certification bodies and (3) philanthropic foundations.

Environmental NGOs

The first group of carriers comprises the environmental NGOs that have created or supported a range of certification initiatives. By creating FSC and exporting the certification model to the fisheries sector, WWF has, of course, been a prominent carrier of the certification model. In addition to creating FSC and MSC, WWF established the Marine Aquarium Council in 1998 to set standards and certify those involved in the collection of ornamental marine life from reef to aquarium. Greenpeace, Friends of the Earth and other environmental NGOs initially pushed for consumer boycotts of certain companies, but they have gradually been enrolled in certification projects, which helped them to coordinate campaigns pushing producers to certify. As discussed in earlier chapters, the collaboration between WWF and more radical NGOs has been effective both in increasing producer adoption of programs and in spreading the certification model.

Certification Bodies

A second type of carrier comprises the certifiers that assess compliance with standards. An entire industry of certifiers, consultants and auditors has emerged around certification programs (see for example Busch 2000; Ponte and Gibbon 2005). Of course, some of these players have been around for a long time, well before the advent of social and environmental certification programs, but the new wave of certification initiatives has presented them with new business opportunities. Certification bodies like the Société Générale de Surveillance (SGS) have a long history of auditing technical standards and are operating commercially to make a profit. Established in 1878 to offer agricultural inspection services to European grain traders, SGS became one of the first certifiers to be accredited by FSC. In 1997, by partaking in the establishment of the labor standards program, Social Accountability International, SGS helped to spread the certification model to the apparel industry (Auld et al. 2007, p. 16). SGS, Scientific Certification Systems and a few other professional certifiers have become accredited to certify operations for nearly every multi-stakeholder social and environmental certification scheme in existence. Table 8.1 provides an overview of the certification bodies that have been accredited to conduct management and/or chain of custody auditing for *both* the FSC and MSC programs. In addition to these certifiers, several specialized certifiers have been accredited for either the FSC program or the MSC program.

Another certifier emerged from an effort by a few activists to address rainforest destruction. In 1987, they founded the Rainforest Alliance as a nonprofit organization dedicated to rainforest protection. Focusing in the beginning only on forests, it went on to become a key player in spreading the certification model to multiple sectors. Forest certification was crucial for its expanding operations, because it was here that it all began. As noted in Chapter 3, the Rainforest Alliance created its SmartWood program in 1989 and helped to create FSC. In 1990, SmartWood certified its first forest operations in Indonesia, expanding operations over the next few years to Brazil, Chile, Mexico, Honduras, Belize, Mozambique and Papua New Guinea (Rainforest Alliance 2007). Since FSC became operational, SmartWood has operated as an accredited third-party certifier. Working with NGOs in Latin America, the Rainforest Alliance soon

Table 8.1 Certification bodies accredited for both FSC and MSC, as of December 2009

	Scope of FSC accreditation		Scope of MSC accreditation	
	Forest management	Chain of custody	Fisheries assessment	Chain of custody
Bureau Veritas Certification	✓	✓	Undergoing accreditation	✓
Control Union Certifications	✓	✓		Undergoing accreditation
Det Norske Veritas Certification	Only in Sweden	✓	✓	✓
Institut für Marktökologie	✓	✓	Undergoing accreditation	✓
Scientific Certification Systems	✓	✓	✓	✓
SGS	✓	✓	✓	✓
TÜV Nord Cert GmbH		✓		✓

Sources: Accreditation Services International (2009) and www.msc.org, accessed December 29 2009.

expanded its certification operations to cover tropical commodities (Auld et al. 2007). In 1994, the first two Chiquita-owned banana farms in Costa Rica were certified, followed the next year by the first coffee farms to be certified in Guatemala (Rainforest Alliance 2007). By 2000, all Chiquita-owned banana farms in Latin America had been certified in accordance with the Rainforest Alliance social and ecological standard. According to the Rainforest Alliance, more that 15 percent of the bananas in international trade currently come from farms it has certified.[1] The certification program now certifies a range of tropical commodities, including cocoa, tea, citrus and cut flowers, and the Rainforest Alliance has partnered with many large corporations to facilitate adoption of the program. In 2007, for example, Unilever announced that by 2015, all tea plantations used for its Lipton tea brand are to be Rainforest Alliance certified.[2]

According to the Rainforest Alliance, its farm certification program results in a number of social and environmental benefits, including decreased water pollution and soil erosion, reduction of pesticides, protection of wildlife habitat and improved conditions for farm workers.[3] Unlike Fair Trade certification, however, it does not guarantee producers a minimum price, nor, according to one observer, does it seem to have a strong impact on working conditions and wages (Conroy 2007, p. 251). Because retailers pay less than the Fair Trade price for certified commodities, Rainforest Alliance certification has been tremendously successful in increasing market adoption, thereby allowing multinational corporations like Chiquita, Unilever and Kraft Foods to capture a large share of the ethical consumer market (Conroy 2007, p. 251).

The Rainforest Alliance was also engaged in early efforts to model a 'Sustainable Tourism Stewardship Council' after FSC, but such a scheme has not yet materialized (Auld et al. 2007, p. 14). The Rainforest Alliance has focused instead on supporting the many local sustainable tourism certification programs in Latin America and helping to create the Sustainable Tourism Certification Network of the Americas. This network works to strengthen sustainable tourism initiatives in the region by identifying best practices, facilitating certification practices, harmonizing systems and sharing information. Following increased attention to carbon sequestration in forests, the Rainforest Alliance moved into the business of carbon-offset verification and validation services, providing carbon-auditing services to forest managers and forest landowners. It has been accredited by the American National Standards Institute to the international ISO standard (ISO 14065) for greenhouse gas validation and verification bodies.[4]

Philanthropic Foundations

A third group of carriers comprises the philanthropic foundations that provide financial support to certification schemes. The emergence of forest certification was significant because it provided them with a project they could jointly support, and demonstrated that certification was a potential solution to a range of environmental and social problems. Bartley (2007) describes the critical role of US philanthropic foundations in the emergence of forest certification. In 1993, a network of foundations, including the Ford Foundation, McArthur Foundation and Rockefeller Brothers Fund, created the Sustainable Forestry Funders network to coordinate funding for the emerging FSC program. These foundations became an important source of funding for a wide range of organizations involved in forest certification, including FSC, its certifiers, buyer groups and environmental groups. According to Bartley, the network provided FSC with financial support, enrolled social movement organizations in certification projects and helped them to coordinate efforts to push producers into certifying. In these ways, 'foundations played a major part in transforming forest certification from a minor experiment to a complex, multinational field' (Bartley 2007, p. 249). Some of the foundations that funded FSC became supporters of MSC; other foundations learned from witnessing the success of forest certification and decided to support the nascent fisheries certification program. As noted in Chapter 6, the Packard Foundation was vital in supporting MSC's transformation from a WWF-Unilever partnership into a fully independent, multi-stakeholder certification program. As is the case with FSC, foundation grants remain MSC's most important source of income. Foundations have also supported a range of other social and environmental certification initiatives, including Fair Trade coffee, Social Accountability International and the Marine Aquarium Council (Auld et al. 2007, p. 17).

FORGING ALLIANCES

Interpersonal ties have been another important factor in the spread of the certification model. Within forestry, relationships among individuals who had been working on community forestry projects in South America and program officers at two US foundations – Homeland and MacArthur – facilitated funding of the early development of FSC and its principles and criteria (Bartley 2007, pp. 239–40). Personal ties among environmentalists from various NGOs also played a valuable role in the process leading to the formation of FSC (Synnott 2005) and MSC (Murphy and Bendell 1997). As noted in Chapter 6, for example, individuals working on marine conservation within WWF learned about FSC from their colleagues, and a member of WWF's forest team was contracted to investigate the possibility of creating a fisheries certification program akin to FSC. In several certification programs, board members are serving or have served on the board of other certification programs (Dingwerth and Pattberg 2009). As discussed below, however, the expansion of certification initiatives meant that strategies to obtain support shifted from personal ties toward formal alliances.

The ISEAL Alliance

In 1999, following the emergence of several certification programs across sectors – often established or supported by the same actors – the relationship among the programs was formalized with the creation of the International Social and Environmental Accreditation and Labeling (ISEAL) Alliance. The eight founding members of the alliance were FSC, MSC, IFOAM, Marine Aquarium Council, International Organic Accreditation Service, Rainforest Alliance, Social Accountability International and Fairtrade Labelling Organizations International.[5] All members are nonprofit organizations, committed to creating credible standards systems. They must demonstrate that their certification or accreditation programs meet *the ISEAL Code of Good Practice for Setting Social and Environmental Standards* (ISEAL Alliance 2006), which specifies general requirements for social and environmental standard setting. The ISEAL Code demands that standards are set in open, transparent and participatory processes. It requires that there must be a 'demonstrable need' to develop the standard, and it specifically demands that 'even the most marginalized stakeholders' must have a voice in the standard-setting process. In short, the Code provides, in its own words, 'a benchmark to assist standard-setting organizations to improve how they develop social and environmental standards' (ISEAL Alliance 2006, p. 1).

ISEAL purports to recognize best-practice standards for social and environmental certification or accreditation. Although the alliance 'does not consider its members to be the only bodies that can legitimately develop environmental and social standards and conformity assessment procedures', it claims that its members are leaders in the field (ISEAL Alliance 2006, p. 1). FSC and MSC are the only certification programs recognized by ISEAL to follow best practice in standard setting within their sectors. In addition to its full members, ISEAL has several associate members that are in the process of meeting its requirement for full membership. The associate membership includes Accreditation Services International (ASI) – the organization created by FSC in 2006 to separate its standard-setting and accreditation functions. As explained in Chapter 6, MSC also decided to outsource accreditation decisions to ASI. It is evident, then, that relationships among non-state standard-setting organizations have become increasingly dense and multifaceted; they are interconnected through interpersonal ties, formal alliances and organizational connections. The formation of ISEAL can be seen as an attempt to promote mimetic processes among its members and solidify a particular organizational model (Dingwerth and Pattberg 2009). By codifying a model for transnational rulemaking, the alliance encourages isomorphism in terms of standard-setting procedures and practices among its members, as well as the adoption of this model by nonmembers. In fact, several nonmembers of the alliance, including the Association for Responsible Mining and the UNCTAD Biotrade Initiative, have used the ISEAL Code to develop social or environmental standards.[6]

Another reason for creating the alliance was undoubtedly to harmonize standards and provide a vehicle for mutual recognition among the member schemes in an effort to gain international recognition of these schemes as facilitators rather than barriers to trade. In the introductory section of its Code, ISEAL states that it aims to promote standards that result in measurable progress toward social and environmental objectives 'without creating unnecessary hurdles to international trade' (ISEAL Alliance 2006, p. 2). WTO provisions speaking directly to such voluntary process and production methods schemes as certification and labeling have been included in the agreements on Technical Barriers to Trade (TBT) and Sanitary and Phytosanitary Measures. Whereas the latter agreement deals primarily with food safety labeling, the TBT agreement can potentially restrict the scope for social and environmental certification programs. The agreement includes, in an annex, a *Code of Good Practice for the Preparation, Adoption and Application of Standards*. To ensure that such labeling programs are nondiscriminatory and do not restrict trade, the TBT Code requires them to consider mutual recognition and, if possible, to base their standards upon existing international standards. The ISEAL Code draws on two government-sanctioned codes: the TBT Code and an ISO Code of good practice for standardization (ISO/IEC Guide 59). This linkage could be seen as a way of establishing the Code as a set of legitimate international guidelines that are compatible with existing standards. There is some evidence to suggest that ISEAL has succeeded in this respect. One measure of success is its membership, which, by mid-2009, had increased from the eight founding organizations to ten full members, nine associate members seeking full endorsement and eight affiliated members that support the alliance. Another measure of success is the increasing number of references to the ISEAL Code as an instrument for recognizing voluntary standards in guidelines issued by governmental and intergovernmental organizations, including FAO, the World Bank and the European Parliament.[7]

PEFC and the International Accreditation Forum

In forestry, the establishment of the PEFC umbrella scheme was motivated not only by the need for a common eco-label for national certification schemes, but also by recognition of the benefits of harmonizing national schemes within a common framework. Again, harmonizing efforts may be seen partly as an attempt to adapt schemes to the international trade regime and reduce the likelihood of complaints from states contending that national schemes are incompatible with WTO rules. The producer-backed programs pursued a different strategy than FSC did in a bid to gain international recognition of their standards. As discussed in Chapter 3, most producer-backed programs based their standards upon intergovernmental criteria and indicators and required certifiers to be accredited by existing government-sanctioned bodies, usually national accreditation organizations. By embedding standard setting and accreditation within governmental and intergovernmental standard systems, they sought to obtain recognition that their labeling programs were not trade restrictive. As noted, another motivation was clearly to enhance the legitimacy of the schemes, compensating for the lack of support from environmental and social stakeholders (Cashore et al. 2004; Bernstein and Cashore 2007).

Both PEFC and the ISEAL Alliance have approached the International Accreditation Forum (IAF) and applied for associate membership (Humphreys 2006, p. 133). Established in 1986, IAF is a worldwide association of national accreditation bodies that has granted associate membership to other international certification and accreditation bodies.[8] In 2002, ISEAL applied for associate status in IAF, but its application was rejected. PEFC, on the other hand, was accepted as an associate member in March 2004 (PEFC 2004a). According to Humphreys (2006, p. 133), the explanation for this outcome may be that both PEFC and IAF are dominated by business interests, and therefore represent the same constituencies. But it is probably equally important that PEFC has succeeded in its strategy of establishing itself as a mutual recognition body aimed at harmonizing national standards and certification rules. As noted, its transition from a European umbrella to a global mutual-recognition body was driven partly by the ambition to include developing countries in the scheme and to not violate WTO rules. Likewise, the primary purpose of IAF is mutual recognition of its accreditation body members in order to 'contribute to the freedom of world trade by eliminating technical barriers to trade'.[9] When PEFC was accepted as an associate member of IAF, the PEFC general secretary announced that: 'The IAF's goal is *Certified once – accepted everywhere*. This goal is fully supported by the PEFC Council'. He added that PEFC would participate in IAF's harmonization processes 'to ensure that accredited forest certification certificates issued in one part of the world are recognized everywhere else in the world' (PEFC 2004a). Admittedly very dry stuff on the face of it, this statement reflects an underlying conflict of interests between industry-backed and NGO-backed certification programs. As explained in Chapter 3, mutual recognition between PEFC and FSC would be ideal for PEFC, whereas it would be detrimental for FSC. Mutual recognition would enable PEFC to communicate to manufacturers and retailers that its standard is equivalent to the FSC standard in terms of stringency and scope. By contrast, mutual recognition would be harmful for FSC, because it would erode FSC's position as the scheme with ultimate recognition from NGOs and the marketplace (see also Cashore et al. 2004). All members of the ISEAL Alliance share the aspiration to set best-practice standards within their sector or industry rather than setting industry-wide standards, accessible to all firms. In this regard, ISEAL and IAF-PEFC represent two very different visions of international standard setting and accreditation.

The World Bank – WWF Alliance

Supporters of certification programs have also formed alliances with international organizations in order to promote certification. One particularly influential alliance was formed as early as 1998, when the World Bank and WWF announced a partnership to promote forest certification and forest protection, particularly in developing countries. The alliance focused on three forest protection targets to be met by the end of 2005: 50 million hectares of new protected areas; 50 million hectares of existing but highly threatened protected areas, to be secured under effective management; and 200 million hectares of production forest under independently certified sustainable forest management (100 million each for developed and developing countries) (Humphreys 2006, p. 171). The World Bank's commitment to certification demanded that it take a clear position on the standards that it would accept. Although the Bank has not formally endorsed any certification scheme, the forest certification requirements specified in its operational policies on forests (World Bank 2002) are remarkably similar to the FSC principles (Humphreys 2006, pp. 173–4).[10] The operational policies document on forests is officially an internal reference guide for World Bank managers, but the Bank can transmit its policy to countries to which it lends, thereby promoting FSC.

In 2004, discussions began on a renewal of the alliance, which was failing to meet its target for certifications (Humphreys 2006, p. 174). Because the World Bank – WWF Alliance wanted to reassess the standards it would accept, World Bank staff approached PEFC, requesting that it participate in a survey to assess forest certification schemes.[11] In a letter to the co-chairs of the alliance, Secretary-General of PEFC, Ben Gunnberg, replied that PEFC declined to participate in the survey because it 'has and continues to have a very strong bias towards the FSC' in its structure, terminology and definitions. According to the secretary-general, the outcome of any use of the survey was therefore 'predetermined'.[12] But he also made it clear that PEFC would like to be involved with the World Bank and WWF in a new and broader alliance that promoted forest certification (see also PEFC 2004b). The alliance was eventually renewed in 2005, with the World Bank and WWF as the only two members (Humphreys 2006, p. 187). By recognizing FSC as a credible forest certification program, the alliance has contributed to its spread – not only in the developing countries to which the World Bank lends, but also in developed countries, where the Bank's support for FSC has surely been noticed. In these ways, international organizations can play an important role in granting – or not granting – legitimacy to non-state certification schemes.

PRACTICING A MULTI-STAKEHOLDER GOVERNANCE MODEL

The founders of multi-stakeholder certification programs often claim that their programs are more inclusive, transparent, democratic and accountable than are many of the formal and informal governance networks in the international and domestic domains (Bernstein and Cashore 2007). These initiatives are also praised by researchers for their transparency, inclusiveness, broad representation of stakeholders and deliberative decision-making processes (for example Oosterveer 2005; Dingwerth 2007; Pattberg 2007). As noted, FSC has served as a model or template for a new type of participatory, multi-stakeholder governance. We have seen that FSC formally balances decision-making power across its environmental, economic and social chambers, and, within each chamber, between the global North and South. In this respect, it is clearly a more inclusive and democratic governance model than that endorsed by the producer-dominated programs. Focusing only on formal organizational structures, however, students of non-state governance schemes can miss critical issues of authority and power that ultimately shape the evolution and practice of multi-stakeholder governance (Boström 2006a). Shifting the focus from formal rules and regulations to practice and implementation, we are left with a critical question: has FSC's multi-stakeholder governance model been able to live up to the high hopes of many stakeholders and observers?

What, then, can be said about FSC's ability to practice the multi-stakeholder ideal over time? Specifically, what kinds of stakeholders have the capacity and resources to influence policy choices and outcomes? One observation is that although FSC formally balances voting power between the global North and South, the proportion of *individual* members as opposed to *organizational* members is much greater in the southern than the northern subdivisions of the chambers. FSC's system of voting parity, regardless of the number of members that join a chamber, means that a vote from a southern organization carries more weight than does a vote from a northern organization (Dingwerth 2007, pp. 169–70). Although FSC's voting rules are favorable to the global South, the imbalanced participation from individual and organizational members from the global North and South reflects more fundamental asymmetries in the power resources of developed and developing countries. Whereas many northern organizations can afford to send representatives to the General Assembly, many of the southern participants are individuals with a travel grant but without the organizational backing that their colleagues from the global North enjoy. Representatives from forest companies, industry associations and other business constituencies in the global North have the knowledge, skills, interpersonal contacts and organizational resources to influence voting outcomes. By contrast, individual members from the global South represent only themselves, often lack networks and alliances and have no access to organizational resources. Rather than changing the asymmetrical power relationship between the global North and South, the current patterns of participation in the General Assembly seem to build on the existing power relationships, unfavorable to the global South.

Turning to another observation, although ultimate decision-making authority rests with the FSC General Assembly, the day-to-day activities of the program are conducted by its international secretariat headed by the executive director. Because the General Assembly convenes only every third year, the secretariat has significant authority and discretion in carrying out both the mandates of the General Assembly and the strategic decisions of the FSC Board. Thus, the secretariat – as well as the executive director appointed by the FSC Board – are influential in shaping the direction of the program for a number of issues. These issues may have a lower profile or be less controversial than some of the issues discussed at the General Assembly, but they may be just as important for the direction being taken by FSC, particularly as they are becoming more and more complex. The consequence of this increased administrative power may be decreased NGO influence within the program. In a study of eco-labeling programs across several sectors, for example, Boström and Klintman (2008) argue that social movement organizations played a more significant role in the early stages of initiating these programs than they did in the subsequent development and governance of the programs.

A third, related observation is that FSC, like any maturing organization, is becoming increasingly complex, routinized and bureaucratic. The entire body of standards, strategies, guidelines and policy documents is growing every year, and for every new issue addressed, new standards must be produced to guide operations in their application and use. FSC's forest management standards now include separate guidance documents focusing on large ownerships, small and low-intensity managed forests, plantations, pesticide use and social issues. Similarly, FSC's chain-of-custody standards include documents on a range of issues, including audits, derogation, outsourcing, product classification, recycling and small operations. There are also standards for group certification, controlled (legally sourced) wood, high conservation value forests, non-timber forest products and national initiatives, as well as trademark rules for on-product labeling, printed materials, promotional use and retailers. In addition to the complexity of the standards documents, the organization itself is becoming more and more complex, with a number of governance levels, governance bodies and working groups addressing various issues. The increasing complexity of both the standards and the organization means that a great deal of knowledge, energy and skills are required to develop FSC, potentially accentuating the variation in influence between strong and weak stakeholders, between the global North and South, and between economic and social/environmental interests.

In general, relationships between actors are likely to change over longer periods as governance arrangements become routinized, bureaucratized and institutionalized, which, in turn, seems to be creating a demand for more and other types of power resources than was initially the case. Preliminary evidence about FSC indicates that, compared to its infancy, the present collaborative arrangements demand more time, money and professional skills to influence policy decisions and outcomes, possibly reducing the ability of NGOs to maintain a strong presence in the program (Boström and Klintman 2008; Boström and Tamm Hallström 2008). Although this development could erode the legitimacy of FSC as a multistakeholder governance program, future research should examine the changing relationships among actors over longer periods.

CHALLENGES FOR SOCIAL AND ENVIRONMENTAL CERTIFICATION PROGRAMS

We have seen that social and environmental certification has grown from a niche market to a globally recognized phenomenon. The growth of forest and fisheries certification during the past decade has created a number of challenges that, at least in the eyes of some observers, threaten to undermine a promising governance tool. The Rainforest Foundation report discussed in Chapter 3 observed that certification has 'evolved from a mechanism needed for effective discriminatory grass-roots boycott campaigns, to become a major international "forest policy tool" embraced by global decision-makers'. According to the report, this change has been accompanied by 'a subtle shift from the use of the FSC principally as a tool for improved forest management to one of improved marketing of forest products' (Counsell and Loraas 2002, p. 14). Similarly, Ward and Phillips (2008, p. 433) observed that with the rapid growth of fisheries certification, 'the links to creation of environmental improvements are being lost amongst the rush to achieve market penetration and advantage'. In an instructive study of the Fair Trade movement, Raynolds and Murray (2007, p. 223) argued that 'the dramatic growth in Fair Trade since 2000 has fueled a number of challenges which threaten to unravel this promising initiative unless its vision and practice can be realigned'. The key challenges facing forest, fisheries, Fair Trade and other social and environmental certification programs are essentially the same, although their objectives vary. Raynolds and Murray (2007) identified four key challenges to the Fair Trade movement, arising from (1) the mainstreaming of Fair Trade distribution, (2) the increasing scale and complexity of Fair Trade production, (3) the challenges of Fair Trade governance and (4) the Fair Trade's shifting movement location. These challenges deserve some detailed attention, because of the parallels to forest and fisheries certification.

The first challenge arises from the pursuit of large-volume markets and business partnerships with large traders, distributors and retailers. According to Raynolds and Murray, the mainstreaming of Fair Trade is engaging supermarket chains such as Tesco, with little visible commitment to social justice principles, and corporations like Nestlé and Chiquita, which are infamous for their exploitative practices in developing countries. The risk is that Fair Trade will amount to little more than a 'clean washing' tool for these companies (Raynolds and Murray 2007, p. 226). The parallel to forest and fisheries certification was the endorsement of this tool from giant companies like Unilever, Tesco, Sainsbury's, Home Depot, IKEA and Walmart. These powerful retailers largely dictate product characteristics, volumes and prices, shifting the costs of certification onto their suppliers. Fears about the erosion of the certification movement's transformative agenda by companies seeking only 'green washing' of their brand names seem justified in this context.

A second challenge is related to the integration of plantations and other large-scale production units into certified production. In the Fair Trade case, this challenge grows out of the mainstreaming of distribution in global commodity chains controlled by powerful retailers demanding high quality, large quantities and uniform product characteristics. According to Raynolds and Murray (2007, p. 227), their case studies demonstrate that the rising quality and volume requirements of major distributors 'are eroding the small farmer base of Fair Trade in the Global South'. As we have seen in this book, forest and fisheries certification programs have a different history than the Fair Trade movement. Whereas the Fair Trade movement historically has a strong commitment to small-scale farmers in developing countries, forest and fisheries certification emerged principally out of environmental concerns. Yet, as with Fair Trade production, one particularly contentious issue within FSC has been the certification of plantations, which, after prolonged debates, was added as a tenth FSC principle in 1996. We have also seen that the favoring of large-scale production and the marginalization of smaller producers and producers in poorer countries have fueled heated debates over forest and fisheries certification. Ironically, we see that forest certification is adopted first and foremost in the regions where, arguably, this tool is needed the least – the global North – because it is easier to adopt in countries where governments set relatively strict environmental and social rules (Cashore et al. 2004; Auld et al. 2007; McDermott et al. 2008). In response to increased concerns over the exclusion of community-based, small-scale and local producers, both MSC and FSC have developed approaches to facilitate certification of small and low-intensity operations in developing countries. MSC has piloted a project to enable small-scale and data-deficient fisheries better access to its label, while FSC has developed specific guidelines for certification of small and low-intensity managed forests. These corrective measures indicate that MSC and FSC are taking the exclusion problems seriously; yet grappling with the social side of certification remains a key challenge for both programs.

The third challenge arises from the changing nature of the certification movement itself. Raynolds and Murray explain that although Fair Trade was initially intended to better the wages and working conditions of producers in developing countries, Fairtrade Labelling Organizations International (FLO) and its national affiliates are now criticized for advancing commercial interests over development interests. FLO has also been criticized for North/South power asymmetries within governance bodies and processes. Similar criticisms have been directed at FSC and MSC. As discussed in this chapter, achieving balanced stakeholder representation can be difficult, as organizations are becoming increasingly complex, routinized and bureaucratic. Concerns have been raised over the limited potential of stakeholders from developing countries to influence rulemaking within the programs. Forest and fisheries certification programs have also been criticized for promoting market adoption at the expense of environmental interests. In the case of forest certification, most environmental NGOs have joined ranks to support FSC and criticize the industry- and landowner-backed programs for trying to co-opt the certification agenda by promoting industry-friendly standards. Yet, their support for FSC is qualified and conditional. In order to keep the support of a broad range of NGOs, FSC must demonstrate that certification makes a difference to on-the-ground practices and forest protection.

A final challenge identified by Raynolds and Murray grows out of the shifting location of Fair Trade within a larger social movement effort to bring social justice and greater equity to international trade. The challenge, as they see, it is to maintain its unique message as part of a movement for an alternative vision of trade and development as it is mainstreamed into global supply chains. Likewise, forest and fisheries certification programs face a key challenge of maintaining a focus on the environmental problems they were created to address, as they are incorporated into global supply chains and mass consumer markets. There is clearly a risk that the demands of giant retailers dilute the environmental and social objectives of certification programs. Such an outcome is likely to result in a withdrawal of support from the social movement actors that created certification programs. It would be detrimental for FSC, which emerged from a broad coalition of stakeholders, but less so for MSC, which emerged from a narrower business-NGO partnership. Arguably, it would not matter for the industry-backed certification programs, which never were supported by a broad range of stakeholders.

The future of certification programs depends upon their handling of these challenges. To address them will not be an easy matter; there are several inherent contradictions among certification objectives that must be carefully considered. As seen in previous chapters, there seems to be an inverse relationship between the stringency of social and environmental standards and producer adoption of these standards. The choice for certification programs is between the opportunity to approve only best practice among industry leaders on the one hand, and the opportunity to create standards that are achievable for a large proportion of the industry's producers on the other. A best-practice benchmark that could be achieved by only a small proportion of producers could deliver real environmental benefits among the few, but would not change industry-wide practices. A less stringent standard that could create a market pull to certify larger numbers of producers runs the risk of amounting to little more than a 'green-washing' tool. Widespread producer adoption of stringent standards seems necessary for effective environmental and social problem solving, but it is difficult to convince or pressure large numbers of producers to adopt stringent standards.

Another related choice exists between the opportunity to make a significant difference in a relatively limited niche market and the opportunity to make a small difference in a mass market. The certification programs discussed in this book have evolved from a niche-market phenomenon to a commercial success in conventional mass markets. In the case of Fair Trade, as noted by Raynolds and Murray, there is a risk that its transformative agenda is being eroded by the market forces it initially set out to transform. The same could be said about forest and fisheries certification and other certification programs. As the unintended consequences of the quest for commercial success have become clear, FSC and MSC have developed specialized programs to address the needs of small producers and marginalized stakeholders, with a particular focus on developing countries. Both programs have also initiated self-assessments to focus on the crucial issue of the extent to which certification delivers tangible environmental and social benefits. Effectively addressing these issues – the needs of smaller producers in poorer countries and the real social and environmental benefits of certification – is likely to be crucial for the credibility and future success of non-state certification programs.

NOTES

1. www.rainforest-alliance.org/agriculture.cfm?id=fruits, August 6, 2009.
2. www.rainforest-alliance.org/news.cfm?id=unilever.
3. www.rainforest-alliance.org/agriculture.cfm?id=main, August 12, 2009.
4. www.rainforest-alliance.org/climate.cfm?id=international_standards, August 6, 2009.
5. www.isealalliance.org/.
6. www.isealalliance.org/index.cfm?fuseaction=Page.viewPage&pageId=498&parentID=490, August 7, 2009.
7. www.isealalliance.org/index.cfm?fuseaction=Page.viewPage&pageId=498&parentID=490, August 7, 2009.
8. See the webpage of the IAF: www.iaf.nu/.
9. www.iaf.nu/index.php?artid=4&place=publnd, August 8, 2009.
10. Paragraphs 10 and 11 of OP 4.36 provide the World Bank's policy on certification. Paragraph 10 bears a distinct resemblance to the FSC Principles.
11. This survey was the Questionnaire for Assessing the Comprehensiveness of Certification Schemes/Systems (QACC).
12. Letter from Ben Gunnberg, PEFC Council Secretary, to David Cassells and Bruce Carbale, co-chairs of the World Bank-WWF Alliance, dated October 7, 2004 – available online at www.pefc.org/internet/resources/5_1184_1054_file.1018.pdf, August 8, 2009.

9. Conclusions

The main contribution of this book is the theoretical and empirical investigation of an as yet under-explored area of contemporary environmental politics: the formation and effectiveness of non-state governance institutions. In this chapter, I review and discuss critical observations from the case studies that help to answer my three overarching research questions:

- How can we explain the emergence of non-state certification programs in the forest and fisheries sectors?
- How do certain program designs emerge, and how and to what extent does program design influence standard-setting outcomes?
- What is the effectiveness of certification programs in resolving or ameliorating the problems that motivated their establishment?

This chapter begins with an examination of the factors that underlie the emergence of forest and fisheries certification programs. The second section reviews evidence that sheds light on the question of how program design influences standard-setting outcomes. The third section examines the organizational mimicry that has occurred among certification programs within and across the forest and fisheries sectors. The fourth section examines the crucial question of what is known about the problem-solving effectiveness of certification. In closing, the fifth section offers some concluding remarks and suggests directions for future research.

INSTITUTIONAL EMERGENCE

Certification programs in the forest and fisheries sectors developed from concerns about environmental degradation, resource depletion and insufficient governmental action to address the problems. Intergovernmental efforts on behalf of forests and fisheries were important for certification initiatives, in what they did and did not produce. Whereas forest certification was a response to the lack of legally binding international rules on forests, fisheries certification emerged to supplement what was perceived by NGOs to be inadequate international rules to address the challenges facing fisheries.

The proliferation of business coordination standards is often seen as being driven by producers who seek to resolve coordination problems, decrease uncertainty and reduce transaction costs. As discussed in Chapter 2, however, there is a fundamental difference between business coordination standards and performance-based certification standards: whereas all producers could benefit from adopting business coordination standards, performance-based standards attempt to ameliorate environmental and social problems that producers otherwise would have little incentive to address. This book demonstrates that most producers decided to adopt FSC's performance-based standards only after intensive NGO campaigns. WWF and other environmental NGOs worked systematically to build coalitions in support of the FSC and to include in the coalitions such powerful retailers as IKEA; Home Depot in Canada and the United States; and British-based B&Q. We have seen that WWF established the first buyer group in the UK to create demand for certified wood as far back as 1991, even before the FSC was up and running. Similar buyer groups were established in a number of other countries. Organized through the Global Forest and Trade Network, buyer groups had considerable success in creating demand for FSC-certified products in Europe and North America.

There was less NGO activism and little direct targeting of producers to persuade them to participate in MSC. Instead, WWF partnered with the major corporation, Unilever, to establish MSC, forging a powerful alliance with one of the world's largest buyers of frozen fish from day one of the scheme's existence. Although the MSC supporters used a less confrontational strategy in creating markets for fisheries certification than FSC supporters did for forest certification, building coalitions and creating alliances in favor of certification projects were crucial in both cases. As both are not-for-profit organizations with small budgets and limited marketing capacity, MSC and FSC were dependent upon alliances with environmental NGOs, retailers and donors. Indeed, in the absence of strategic bargaining positions within well-established producer and supply-chain networks, support from environmental NGOs and strategic alliances with powerful retailers were essential to convince producers to sign up to the schemes. Consumers' actual buying behavior or willingness to pay a price premium for eco-labeled products was less important for the emergence of forest and fisheries certification schemes. Nonetheless, environmental groups would certainly have had less success in their efforts to create markets for eco-labeling without the threat of consumer boycotts or the hope of price premiums or greater market access.

The size, ownership and export dependence of an operation affected its vulnerability to NGO targeting. Variation in *forest industry structure* emerged as a particularly significant variable for explaining divergent forest certification choices in Sweden and Norway. Whereas the large, export-dependent Swedish forest companies responded to advocacy group and market pressures by adopting the relatively stringent FSC standards, nonindustrial forest owners in both Norway and Sweden rejected this scheme because of narrower market exposure and their belief that the FSC standards were unsuited for certification of small-scale nonindustrial forestry. The nonindustrial forest owners responded collectively to NGO pressure to adopt the FSC standards by creating landowner-dominated schemes with more discretionary and flexible standards. Their strong associational systems facilitated collective and strategic responses to NGO pressure to certify.

The processes investigated here also show that path dependencies occur and create effects that shape, constrain and limit future policy choices (cf. George and Bennett 2004; Pierson 2004). As Cashore et al. (2004) have argued, certification choices at critical junctures create 'lock-in effects' (Pierson 1993) that constrain future choices and increase the costs of changing course. It was not predetermined that Swedish forest companies would choose FSC certification merely because they were dependent on export markets and were exposed to NGO pressures to certify. In fact, a number of forest companies in other countries – comparable to the Swedish companies in size and dependence on paper and wood products exports to environmentally sensitive markets – rejected this program and worked instead to create industry-dominated schemes (Cashore et al. 2004).

Similarly, small-scale forest owners in Sweden and Norway were not predestined to reject FSC certification merely because they were less exposed to NGO targeting and market pressure than the big forest companies were. Indeed, all six regional Swedish forest owner associations agreed collectively to join the FSC working group and were close to accepting the proposed FSC standards. In the end, however, they withdrew from the working group, largely over disagreement with Sami representatives about reindeer herding rights on private forestland. This decision paved the way for the creation of a landowner-dominated scheme in Sweden (Cashore et al. 2004). If the forest owner associations had decided to remain on the FSC working group and continue the negotiations with Sami representatives and the other stakeholders, there might have been only FSC-certified forestland in Sweden today rather than two competing schemes.

My argument is not that structural variables were unimportant for certification outcomes; the size, ownership and export dependence of an operation clearly influenced certification choices. Rather, I want to stress that structural variables did not fully determine certification outcomes. Standard setting is a bargaining process and the outcome is a result of framing activities, power struggles and competition for influence among various stakeholders. We have seen that the strategies and actions of standard setters influence the way the standard-setting process unfolds. In Sweden, WWF initiated a FSC working group, worked systematically to create a coalition in support of FSC certification and succeeded in persuading the big forest companies and the forest owner associations to participate in the working group. By contrast, the forest owner associations initiated the Norwegian Living Forests project and assumed leadership in the standard-setting process. These associations had the upper hand in Norway, and NGO efforts to convince forest owners of the benefits of FSC certification never succeeded.

THE UNFOLDING AND OUTCOME OF STANDARD-SETTING PROCESSES

We have seen that different interest groups struggled intensively to craft the decision-making rules and procedures of non-state governance schemes, indicating that there is a belief within interest groups that *constitutive* rules define the space for influencing the *regulative* rules being produced (cf. Pattberg 2007; Boström and Klintman 2008). Their belief is, in other words, that the organization of rulemaking processes makes a significant difference to rulemaking outcomes. But does organizational form really matter in the sense of influencing standard-setting processes and outcomes?

A key finding from the case studies is that inclusiveness in standard development and governance bodies enhances the legitimacy and rulemaking authority of non-state governance institutions. The legitimacy of non-state governance schemes is largely determined by the evaluations of environmental NGOs, producers, retailers and consumers (Cashore et al. 2004). Because of the 'symbolic capital' of environmental NGOs, their support is vital for market-based certification schemes (Boström 2006a). Environmental NGOs could be seen as granting legitimacy to certification schemes in exchange for participation in the standard-setting process and, ultimately, for influence on producer behavior (Pattberg 2007). Similarly, participation by producer associations could create a sense of ownership of standard-setting outcomes and enhance the legitimacy of a scheme among producers.

In the case of FSC, participation by a broad range of environmental, social and economic stakeholders enhanced the legitimacy of the program among retailers and consumers, thus increasing supply-chain support for certification. Inclusiveness in decision-making processes also helped to enhance collaboration and problem-solving efforts among stakeholders with different interests. However, stakeholders who felt deprived of real decision-making power or who were unwilling to compromise with other participants left the standard-setting groups. In Norway, after having participated in the Living Forest standard development group, Friends of the Earth declared a few years later that they no longer supported the certification scheme. The environmentalists claimed that the forest owners would not compromise on key environmental issues in the interpretation of the standard. Balancing the formal decision-making powers and rights of various stakeholders appears to be crucial in non-state governance schemes. If business interests dominate rulemaking at the expense of other stakeholders, environmental and social movement groups are not likely to support the scheme. Conversely, if environmental and social movement groups dominate rulemaking, producers are unlikely to participate and implement the rules on a voluntary basis.

As expected, the empirical evidence shows that environmental standards are more likely to be stringent when environmental NGOs are systematically included in standard-setting processes. Conversely, the standards are more likely to be discretionary and flexible when organizational arrangements favor business actors at the expense of other stakeholders. In FSC, WWF and other NGOs deliberately designed organizational arrangements and procedures to eliminate business dominance and to encourage collaborative rulemaking. Within FSC, the environmental and social chambers (comprising two-thirds of the votes in the General Assembly) can always veto proposals they do not support. With only one-third of the votes in the FSC General Assembly, the economic chamber cannot dominate rulemaking in the scheme. By contrast, forest industries and landowners generally dominate rulemaking and governance in FSC-competitor programs. As noted in Chapter 3, voting rights in the PEFC Council are based on the size of the forest owners' land; environmental and social stakeholders have no formal voting rights. These different organizational forms have resulted in different types of standards and certification requirements. Whereas FSC certification is generally based on prescriptive, performance-based standards, PEFC-endorsed schemes place greater weight on standards of procedure, organizational and management measures, and flexibility in applying the standards (Cashore et al. 2004).

The upshot is that initiators of standard-setting programs must carefully consider the type of organizational arrangements that are most suitable for achieving their objectives. In general, including a broad range of stakeholders such as environmental and social NGOs in standard-setting projects is likely to enhance the legitimacy of the programs, but it is also likely to result in relatively stringent standards. Business domination is more likely to result in discretionary and flexible standards, but environmental and social stakeholders are less likely to support the standards and lend credibility to the scheme. On the other hand, we have seen that producers do not always share identical interests. If consumers and retailers value environmentally responsible practices, environmental frontrunners in the business community could benefit from the adoption of stringent standards. Frontrunners can adopt certification standards without having to undertake costly management and behavioral changes, a position that obviously provides them with a competitive advantage vis-à-vis producers who would have to undertake costly reforms in order to become certified. First movers can also shape the rules to match their technical and operational capacities, resulting in higher switching costs for late movers (Mattli and Büthe 2003). In Sweden, for example, the forest companies clearly regarded FSC certification as a competitive advantage when dealing with environmentally concerned export markets such as the UK and Germany. During the 1980s, as a result of such environmental reforms in the forest companies as the development of environmental management plans, the hiring of ecologists and the education of personnel in ecology and environmental protection, they were well prepared for the adoption of the relatively stringent FSC certification standards. Whereas the Swedish forest companies could benefit from environmental preparedness and operations of scale, transaction costs for private forest owners practicing small-scale forestry would have been much higher.

Similarly, various environmental NGOs do not always share the same interests, and sometimes disagree on strategies or objectives. FSC is both a site of and a source of occasional conflict within the environmental community (Bartley 2007). Within FSC, there is an ongoing debate between WWF – an enthusiastic and pragmatic supporter of the scheme – and more radical groups such as Greenpeace and the Rainforest Action Network. WWF would like to see FSC develop into the world's largest certification scheme, and has stressed the need for some flexibility to accommodate business interests. By contrast, Greenpeace and other environmental NGOs have argued for more stringent certification requirements, maintaining that FSC should be an exclusive scheme, in which only the best companies can participate. Conflict levels were highest at FSC's inception and have since abated somewhat, but the differing views on what the FSC should be and how it should develop are reflected in ongoing discussions within its council, General Assembly and other governing bodies.

In sum, we see that there can be divergent interests not only among the various stakeholders, but also within the environmental community and the business community. It is evident, then, that when explaining the outcome of standard-setting processes, one must consider the *configuration* of interests within standard development coalitions. The case studies show that the greatest likelihood of business–environmental NGO agreement on relatively stringent standards occurs when there are producers who could benefit from adopting such standards. Given the credibility that environmental NGOs lend to the more stringent certification schemes, environmental frontrunners on the producer side tend to favor participation in such schemes. By contrast, those who favor more lenient and discretionary standards tend to prefer participation in producer-dominated schemes that give them smaller adoption costs and greater influence over standard-setting outcomes.

ORGANIZATIONAL MIMICRY

It is important to recognize that standard setting is neither an isolated event nor a process with a final outcome (Auld et al. 2008). Standards are always negotiated and implemented in a specific context and standard setting is an iterative process involving adjustment, adaptation and renegotiation of standards in light of new concerns, demands and knowledge. We have seen that competition between NGO-backed and producer-backed schemes influences standard-setting processes. The Swedish FSC standard was initially more stringent and prescriptive than the PEFC-endorsed standard created by the forest owner associations. Given the competition for credibility and support from stakeholders, however, the two schemes have adjusted their standards and become more similar. Whereas FSC has adjusted its rules to better accommodate the needs of forest companies, PEFC has changed 'upward' in an effort to enhance credibility among environmental NGOs and in the marketplace (Cashore et al. 2004).

The Swedish case indicates that the difference between competing standards cannot be too great. If a particular standard becomes too stringent, most producers will simply choose to participate in a competing standard. But if the competing standard is too discretionary and lenient, it is not likely to be supported by NGOs or the marketplace. As a result, competing standards are likely to influence one another and the space for making mutual adjustments. There is, however, a key difference between NGO-backed schemes like FSC and producer-dominated programs. Whereas FSC needs to demonstrate that it is 'best in class' in order to retain environmental NGO support and credibility, producer-dominated schemes often need merely to convince important buyers that their labels are 'better than average' or better than non-labeled products (cf. Cashore et al. 2004; Boström and Klintman 2008). Because producer-dominated certification programs are dependent upon support from industry players and supply-chain actors, but not necessarily upon widespread support from environmentalists, they do not have to be more stringent than competing schemes. The motivation for creating producer-led programs was, after all, to create a more industry-friendly alternative to FSC. By contrast, the legitimacy of FSC rests on its being the most environmentally stringent and demanding certification program in the forest sector.

Notwithstanding the two different certification models, producer-dominated schemes have imitated some of the organizational arrangements in the FSC-style model. We have seen that producer-backed schemes have constituted themselves more or less independently of the producer associations that established them. The have also become increasingly open to participation from stakeholders outside the forestry community. Over time, we see evidence of increasing organizational homogeneity within the certification field. Recall from Chapter 2 that homogenization within organizational fields is a result of coercive isomorphism, mimetic processes or normative pressures (DiMaggio and Powell 1983). In the cases of forest and fisheries certification, we have observed all three processes at work. Coercive isomorphism has resulted from pressures from environmental NGOs and preferences for particular organizational forms from donors, charities and philanthropic foundations. The preferences of funding bodies for FSC and MSC have been crucial to the growth of these organizations and to the construction of a certification organizational field (cf. Bartley 2007). Mimetic processes occur when a number of organizations imitate a specific organizational model that is considered to be particularly legitimate and successful. We have seen that the success of FSC in attracting widespread support among market players and NGOs has helped to spread the FSC-style governance model, which, in turn, is legitimated by widely held norms and beliefs about appropriate ways of organizing rulemaking and governance in modern society. In addition, normative pressures occur as professionals occupy similar positions across a wide range of organizations and introduce their occupational principles, norms and values in those organizations. This process is perhaps most apparent in the entire sub-sector of auditing activities that did not previously exist. Certification bodies that audit forests and fisheries are occupied by professional auditors with similar educational backgrounds and value systems. These auditors introduce their occupational principles and practices in certification bodies, whether they audit the operations of producers certified by MSC in the fisheries sector or FSC and producer-backed programs in the forest sector. We can observe, then, that certification schemes are embedded in particular organizational fields and molded by institutionalized norms and values in those fields.

In early neoinstitutional work, organizations are said to reflect, but never to transform, institutionalized norms and values in the environments in which they are situated. From this perspective, formal structure is seen as 'myth and ceremony' (Meyer and Rowan 1977) that merely tend to reproduce overarching metanorms and powerful value orientations. My case studies support more recent institutional work demonstrating that organizations transform institutionalized norms and innovate to create institutional change (for example Sahlin-Andersson and Engwall 2002). The chapters on fisheries certification focus on knowledge acquired from certification experience gleaned within the forest sector – an understanding that paved the way for MSC. It was through this process that MSC mimicked some of the features of FSC, while strategically avoiding other features. This selective mimicry resulted in a more streamlined approach to governance and stakeholder involvement. The initiators of MSC decided against an open-membership organizational model. Ultimate decision-making authority was granted instead to the appointed Board of Trustees, comprising members from industry, environmental NGOs, the scientific community and the seafood retailers. The Stakeholder Council advises the Board of Trustees, but it cannot veto or overrule decisions made by the Board. MSC requires fisheries seeking certification to comply with substantive performance requirements; but its standards are narrower than FSC's standards, and exclude social issues such as the needs of local communities and the rights of indigenous peoples and workers. It also differs from FSC in its decision to avoid delegating authority to make a global standard locally appropriate to national affiliates. MSC's governance structure, standards and certification requirements have courted controversy about the extent to which MSC has empowered environmental and social groups with a stake in fisheries governance.

As discussed in Chapter 3, although producer-backed forest certification schemes have mimicked the FSC-style organizational model by enhancing their autonomy and openness to other stakeholders, they have acted strategically to maintain control of the standard-setting process. Rather than passively absorbing popular organizational recipes, they have adapted selectively to institutionalized norms and values by adopting certain recipes while carefully filtering out the management prescriptions of which they did not approve. Indeed, producer-backed and NGO-backed schemes are struggling to craft the appropriate norms, rights, rules and decision-making procedures in the certification organizational field. Whereas environmental NGOs typically have invoked norms and beliefs about stakeholder democracy, deliberation and transparency in institution-building processes, forest owners have invoked norms and value orientations related to their sense of independence and identity as stewards of their forests. Institution building should be seen, then, as a struggle between competing sets of norms and value orientations, with no predetermined outcome (cf. Conca 2006). Non-state governance institutions developed in part through collaboration between environmentalists and producers and in part through contestation among actors who wanted the institutions to serve to advance their values, beliefs or interests.

In conclusion, the empirical material shows that institutional environments influence but do not fully determine formal structure in certification schemes. There is scope for agency and transformation of underlying normative structures in the certification field. We have seen that certification schemes have causal autonomy and that they are not merely a reflection of configurations of power and interests among stakeholders or broader social orders. On the other hand, it is important to recognize that organizational recipes and institutionalized norms and beliefs limit the range of available options in the certification field, requiring standard setters to choose among a limited range of acceptable or appropriate organizational forms.

PROBLEM-SOLVING EFFECTIVENESS

Because FSC seeks to provide stricter and more demanding forest management rules than those agreed upon by governments, its principles are not explicitly linked to any intergovernmental forest principles or rules. In contrast, we have seen that the MSC standards are based upon the 1995 FAO Code of Conduct for Responsible Fisheries. This is clearly an attempt to reassure governments that MSC does not seek to establish a competing non-state regime to the elaborate international fishery regime, centered on the 1982 Law of the Sea Convention. A wide range of multilateral, regional and bilateral fishery treaties, as well as international soft law (such as the FAO Code of Conduct), supplement the ocean law codified in the Law of the Sea Convention. MSC operates within this regulatory framework.

The divergent roles of FSC and MSC are related to different ways in which forests and fisheries are governed. Forests are national resources governed by domestic authorities and owners. Although states own three-quarters of the world's forests, most governments have transferred management authority to private companies through logging concessions. National laws regulate access to and use of forest, but forest companies and private owners are often given great leeway to exploit forestland, and lack of forest law enforcement remains a major problem, particularly in developing countries in the tropical zone. Forest certification schemes could directly influence the way forest companies and landowners manage forests and conduct logging operations.

In contrast, marine fish stocks are common pool resources managed by governments through international and regional collaboration arrangements, and there is little scope for private authorities like MSC to influence fisheries management regulations. Whereas FSC seeks to establish a global standard for well-managed forests in the absence of a global forest convention, MSC essentially aims to improve fisheries management practices through setting standards that supplement multilateral and regional fisheries agreements. Because most fisheries are under the control of governments, fish stocks require government intervention for their conservation.

Certification bodies may identify regulations that need to be changed to allow for certification, but it is not their task to appeal to governments to change management regulations. Rather, the applicant (fishing industry or other stakeholders) may work with government regulators to change regulatory frameworks in ways that would allow certification of fisheries that do not meet MSC standards (Leadbitter et al. 2006). Likewise, if governments believe that certification is vital for the economic viability and market access of the fishing industry, they may take the initiative to change management rules to allow for the certification of fisheries. Governments also have the power to change the course of certification initiatives. Indeed, through the development of fisheries eco-labeling guidelines within the FAO, governments have taken steps to regulate certification initiatives in the fisheries sector. Thus, it is clear that states with a significant stake in fisheries governance are not willing to leave the creation of certification rules and procedures completely to the discretion of non-state actors, and they are able to regain some control over non-state rulemaking.

The global scale of overfishing and depleting fish stocks is a significant challenge to certification as a tool for addressing problems that are rarely contained within a single fishery. Many fish stocks are straddling and highly migratory, and there are often multiple access rights to shared fish resources. Certifying one fishery at a time cannot resolve large-scale fisheries management problems that require intergovernmental efforts to address. Although process improvements in MSC-certified fisheries indicate that certification can benefit fisheries management and practices, overfishing and the depletion of fish stocks continue, largely unabated.

Looking at evidence of behavioral changes following forest certification, we see that forest companies and landowners that certify have had to change their management operations. Studies of CARs issued by certifiers show significant attention being paid to improvements in internal monitoring and auditing in forest organizations. These studies also indicate that forest organizations have had to attend to ecological aspects of their management more carefully following certification (see Auld et al. 2008). It seems to be a warranted conclusion, then, that certification has resulted in changes in on-the-ground management. Yet, the environmental impact of certification in many countries seems to have been quite marginal. As in the case of fisheries certification, issues of scale represent a significant hurdle in using certification as a tool to address environmental problems that are rarely contained within a single forest, such as the management of larger predators requiring millions of hectares of contiguous habitats and forest protection at the landscape level (Auld et al. 2008, p. 199). Protection on an individually certified tract can lead to higher pressure for extraction on noncertified lands. In regard to reducing pressure for deforestation, there is also broad recognition that certification provides an inadequate counterbalance to greater economic incentives for land-use conversion (Gullison 2003; Auld et al. 2008, p. 199). These observations lend support to Peter Dauvergne's worrying conclusion, in his study of the consequences of consumption for the global environment, that:

> The globalization of environmentalism is improving management on *some* measures, significantly decreasing the per unit impacts of *some* consumer goods for *some* consumers. But it's failing to prevent the globalization of investment, trade, and financing – powered by multinational corporations and strong states – from displacing a disproportionate share of the ecological costs of rising consumption into the most fragile ecosystems, onto the poorest people, and into distant times. (Dauvergne 2008, p. 231; emphasis in original)

On balance, although certification seem to change *some* management practices and create better environmental outcomes in *some* cases, it does not seem to be an effective environmental institution in the sense of addressing some of the most serious environmental challenges in the forest and fisheries sectors. That said, we still know too little about the environmental impact and efficacy of certification as a problem-solving instrument. Neither do we have evidence about variation in the impact of different certification programs. These are areas in urgent need of closer examination.

Turning to adoption patterns, we have seen that there are challenges related to self-selection in voluntary certification programs; when standards are high, not all companies and landowners are willing or have the capacity to participate. In order to be effective, certification programs need participation from a critical mass of producers. Participation from a few industry leaders could set an example for the rest of the industry, but if most producers reject certification, there would be no broad-scale change of industry-wide practices. Fisheries certification has proliferated among well-managed fisheries in developed countries, but not in developing countries. Similarly, an examination of adoption patterns around the world shows that certified forestlands are skewed in favor of temperate and boreal forests, indicating that certification has spread primarily among producers who face relatively low adoption costs. Patterns of adoption also shows that producer-backed schemes have outperformed the FSC in many countries and regions; by the end of 2009, they had certified about twice as much forestland as that certified by the FSC. The wider producer acceptance of the PEFC is an indication that producers tend to prefer participation in schemes with less stringent and prescriptive standards than FSC offers. But the character of the forest operation also influences adoption choices. Patterns of adoption indicate that whereas nonindustrial owners may reject relatively stringent standards because of the high fixed costs of preparing for and responding to certification audits, large companies can afford to participate because of the benefits of economies of scale. Accordingly, certification schemes do have consequences – such as favoring large-scale over small-scale production – that were not intended or anticipated by those who created these schemes. Although both FSC and MSC have introduced specialized programs to reduce entry barriers for small producers from developing countries, patterns of adoption continue to raise questions about the effectiveness of certification.

In sum, there is clearly a dilemma in setting stringent standards that would compel producers to undertake reforms they otherwise would not pursue, while simultaneously ensuring broad-scale participation. Although a few certification frontrunners could adopt stringent standards at a relatively low cost and obtain a competitive advantage in markets that value certified wood, the majority of producers will have to be convinced of the benefits of participation or coerced into adopting standards by activist targeting and campaigns. Regarding the underrepresentation of developing country producers in certification programs, more research is needed on the economic, political and social factors that facilitate or hinder the spread of certification initiatives in developing countries.

CONCLUDING REMARKS

Certification programs have emerged in recent years to become innovative and dynamic institutions for non-state governance. Sometimes referred to as 'governance without government' (cf. Rosenau and Czempiel 1992), the emergence of such programs has been taken as evidence supporting the claims that the state is less powerful than it has been in the past and that authority is being relocated from public to private institutions. The evidence presented in this book, however, supports the position that states, through the regulatory system and the political and administrative culture, have influenced non-state rulemaking initiatives and encouraged private actors to participate in certification programs. By focusing predominantly on market dynamics and the strategies of non-state actors, many analysts of non-state governance schemes tend to ignore or downplay the role of political and regulatory frameworks in the implementation of these programs. Although certification programs emerged in response to inadequate intergovernmental regulations, analysts would be well advised to pay attention to the legal, socioeconomic and political contexts that facilitate or hinder successful implementation. There is a great deal of evidence to suggest that effective implementation of certification programs requires well-functioning legal systems, property rights and national and local administrations that work. Non-state certification initiatives cannot fully supplant domestic legislation and its enforcement by public authorities. In fact, their successful functioning seems, in part, to depend on such legislation and enforcement. Certification may do little, therefore, to improve the overall protection of forests, fisheries or other natural resources in countries or regions where governmental institutions, legislative frameworks and law enforcement mechanisms are weak.

It is critical to recognize, then, that private and public rulemaking processes are closely intertwined; that private regulatory regimes influence public regulatory regimes, and vice versa; and that the absence of one affects the dynamics in the other. The process of private and public institution building is part of a broader effort to address collective problem complexes. More research is needed on the role of certification as an integral part of governmental, intergovernmental and civil society efforts to resolve some of the most pressing global environmental problems facing humankind today; overfishing, forest degradation, land-use change and loss of biodiversity.

References

Accreditation Services International (2009), *Accredited Certification Bodies for the Forest Stewardship Council (FSC) Program*, 2 October, Bonn: Accreditation Services International.

Adler, E. (1997), 'Seizing the middle ground: constructivism in world politics', *European Journal of International Relations*, **3** (3), 319–63.

AF&PA (American Forest and Paper Association) (2004), *Sustainable Forestry Initiative 2005–2009 Standard*, Washington, DC: AF&PA.

Agnew, David (2008), 'Case study 1: toothfish – an MSC certified fishery', in Trevor Ward and Bruce Phillips (eds), *Seafood Ecolabelling: Principles and Practice*, Oxford: Wiley-Blackwell, pp. 247–58.

Agnew, David, Chris Grieve, Pia Orr, Graeme Parkes and Nola Barker (2006), *Environmental Benefits Resulting from Certification Against MSC's Principles and Criteria for Sustainable Fishing*, London: MRAG UK and Marine Stewardship Council.

Allison, E.H. (2001), 'Big laws, small catches: global ocean governance and the fisheries crisis', *Journal of International Development*, **13**, 933–50.

Andersson, M. (2002), *Naturhänsyn på certifierade private skogsfastigheter – en jämförelse i praktiken mellan FSC och PEFC i Sydsverige* [*Nature Conservation on Certified Private Forestland – a Comparison Between the Two Certification Systems FSC and PEFC in Southern Sweden*], Alnarp, Sweden: Sveriges lantbruksuniversitet.

Auld, G. (2007), 'The origins and growth of social and environmental certification programs in the fisheries sector', paper presented at the 11th Annual Conference of the International Society for New Institutional Economics, June 21-3, Reykjavik.

Auld, G. (2009), *Reversal of Fortune: How Early Choices Can Alter the Logic of Market-Based Authority*, Ph.D. dissertation, New Haven, CT: Yale University School of Forestry and Environmental Studies.

Auld, G., C. Balboa, T. Bartley, B. Cashore and K. Levin (2007), 'The spread of the certification model: understanding the evolution of non-state market driven governance', paper presented at the 48th International Studies Association Convention, Chicago, IL.

Auld, G., L.H. Gulbrandsen and C.L. McDermott (2008), 'Certification schemes and the impacts on forests and forestry', *Annual Review of Environment and Resources*, **33**, 187–211.

Auld, G. and L.H. Gulbrandsen (2010), 'Transparency in non-state certification: consequences for accountability and legitimacy', *Global Environmental Politics*, **10** (3).

Aulén, Gustaf and Stefan Bleckert (2001), *Skogsduvan* (The Stockdove), Report commissioned by WWF Sweden, Swedish Society for Nature Conservation, Swedish Federation of Forest Owners and Swedish Forest Industries Federation.

Axelrod, Robert (1984), *The Evolution of Cooperation*, New York: Basic Books.

Bache, Ian and Matthew Flinders (eds) (2004), *Multi-level Governance*, Oxford: Oxford University Press.

Balton, D. (1996), 'Strengthening the law of the sea: the new agreement on straddling fish stocks and highly migratory fish stocks', *Ocean Development and International Law*, **27** (1–2), 125–51.

Barnett, M. and M. Finnemore (1999), 'The politics, power, and pathologies of international organizations', *International Organization*, **53** (4), 699–732.

Barnett, Michael and Martha Finnemore (2004), *Rules for the World: International Organizations in Global Politics*, Ithaca, NY: Cornell University Press.

Barrett, Scott (2003), *Environment and Statecraft: The Strategy of Environmental Treaty-making*, Oxford: Oxford University Press.

Bartley, T. (2003), 'Certifying forests and factories: states, social movements, and the rise of private regulation in the apparel and forest products fields', *Politics and Society*, **31** (3), 433–64.

Bartley, T. (2005), 'Corporate accountability and the privatization of labor standards: struggles over codes of conduct in the apparel industry', *Research in Political Sociology*, **14**, 211–44.

Bartley, T. (2007), 'How foundations shape social movements: the construction of an organizational field and the rise of forest certification', *Social Problems*, **54** (3), 229–55.

Bass, Stephen, Kirsti Thornber, Matthew Markopoulos, Sarah Roberts and Maryanne Grieg-Gran (2001), *Certification's Impact on Forests, Stakeholders and Supply Chains*, Nottingham: International Institute for Environment and Development.

Bennett E. (2001), 'Timber certification: where's the voice of the biologist?', *Conservation Biology*, **15**, 308–10.

Bernstein, Steven and Benjamin Cashore (2004), 'Nonstate global governance: is forest certification a legitimate alternative to a global forest convention?', in John Kirton and Michael Trebilcock (eds), *Hard Choices, Soft Law: Voluntary Standards in Global Trade, Environment and Social Governance*, Aldershot: Ashgate, pp. 33–64.

Bernstein, S. and B. Cashore (2007), 'Can non-state governance be legitimate? An analytical framework', *Regulation and Governance*, **1**, 347–71.

Boli, J. and G.M. Thomas (1999), 'World culture in the world polity: a century of international non-governmental organization', *American Sociological Review*, **62** (2), 171–90.

Boström, M. (2003), 'How state-dependent is a non-state-driven rule-making project? The case of forest certification in Sweden', *Journal of Environmental Policy and Planning*, **5** (2), 165–80.

Boström, M. (2006a), 'Regulatory credibility and authority through inclusiveness: standardization organizations in cases of eco-labeling', *Organization*, **13** (3), 345–67.

Boström, M. (2006b), 'Establishing credibility: practicing standard-setting ideals in a Swedish seafood-labelling case', *Journal of Environmental Policy and Planning*, **8** (2), 135–58.

Boström, Magnus and Christina Garsten (eds) (2008), *Organizing Transnational Accountability*, Cheltenham, UK and Northampton, MA, USA: Edward Elgar.

Boström, Magnus and Mikael Klintman (2008), *Eco-standards, Product Labelling and Green Consumerism*, Basingstoke: Palgrave Macmillan.

Boström, M. and K. Tamm Hallström (2008), 'NGO participation in global social and environmental standard setting', paper presented for the ISA World Forum of Sociology, September 5-8, Barcelona.

Breitmeier, Helmut, Oran R. Young and Michael Zürn (2006), *Analyzing International Environmental Regimes: From Case Study to Database*, Cambridge, MA: MIT Press.

Bridgespan Group (2004), *Fishery Certification: Summary of Analysis and Recommendations*, Boston, MA: Bridgespan Group.

Brunsson, Nils and Bengt Jacobsson (eds) (2000), *A World of Standards*, Oxford: Oxford University Press.

Burgmans, A. (2003), 'Cooperation is catching', *Our Planet The Magazine of the United Nations Environment Programme*, **13** (4), 22–3.

Busch, L. (2000), 'The moral economy of grades and standards', *Journal of Rural Studies*, **16**, 273–83.

Carr, C.J. and H.N. Scheiber (2002), 'Dealing with a resource crisis: regulatory regimes for managing the world's marine fisheries', University of California International and Area Studies Digital Collection, **1**, accessed September 17, 2009 at http://repositories.cdlib.org/uciaspubs/editedvolumes/1/3.

Cashore, B. (2002), 'Legitimacy and the privatization of environmental governance: how non-state market-driven (NSMD) governance systems gain rule-making authority', *Governance: An International Journal of Policy, Administration, and Institutions*, **15** (4), 503–29.

Cashore, Benjamin, Graeme Auld and Deanna Newsom (2004), *Governing through Markets: Forest Certification and the Emergence of Non-State Authority*, New Haven, CT: Yale University Press.

Cashore, B., G. Auld, S. Bernstein and C. McDermott (2007a), 'Can non-state governance "ratchet up" global environmental standards? Lessons from the forest sector', *Review of European Community and International Environmental Law*, **16** (2), 158–72.

Cashore, B., E. Egan, G. Auld and D. Newsom (2007b), 'Revisiting theories of nonstate market driven (NSMD) governance: lessons from the Finnish forest certification experience', *Global Environmental Politics*, **7** (1), 1–44.

CEPI (Confederation of European Paper Industries) (2001), *Comparison Matrix of Forest Certification Schemes*, Brussels: CEPI.

Chayes, Abram and Antonia Handler Chayes (1995), *The New Sovereignty: Compliance with International Regulatory Agreements*, Cambridge, MA: Harvard University Press.

Checkel, J.T. (1997), 'International norms and domestic politics: bridging the rationalist-constructivist divide', *European Journal of International Relations*, **3** (4), 473–95.

Checkel, Jeffrey T. (ed.) (2007), *International Institutions and Socialization in Europe*, Cambridge: Cambridge University Press.

Clapp, J. (1998), 'The privatisation of global environmental governance: ISO 14000 and the developing world', *Global Governance*, **4** (3), 295–316.

Coglianese, C. and J. Nash (2002), 'Policy options for improving environmental management in the private sector', *Environment*, **44** (9), 10–23.

Conca, Ken (2006), *Governing Water: Contentious Transnational Politics and Global Institution Building*, Cambridge, MA: MIT Press.

Conroy, Micael E. (2007), *Branded! How the 'Certification Revolution' is Transforming Global Corporations*, Gabriola Island, BC: New Society Publishers.

Constance, D.H. and A. Bonanno (2000), 'Regulating the global fisheries: the World Wildlife Fund, Unilever, and the Marine Stewardship Council', *Agriculture and Human Values*, **17**, 125–39.

Counsell, Simon and Kim Terje Loraas (2002), *Trading in Credibility: The Myth and Reality of the Forest Stewardship Council*, London: The Rainforest Foundation.

CPET (Central Point of Expertise on Timber) (2004), *UK Government Timber Procurement Policy: Assesment of Five Forest Certification Schemes*, CPET Phase 1 final report, November, London: CPET.

CPET (2006), *Evaluation of Category A Evidence: Review of Forest Certification Schemes: Results*, December, London: CPET.

Cutler, A. Claire, Virginia Haufler and Tony Porter (eds) (1999), *Private Authority and International Affairs*, Albany, NY: SUNY Press.

Dahl, Lena (2001), *FSC i praktiken* [FSC in Practice], Stockholm: Swedish Society for Nature Conservation and WWF.

Dahl, Lena (2002), *Bakom kulisserna: En analys av PEFC i Sverige i 2002* [*Behind the Scenes: An Analysis of PEFC in Sweden in 2002*], Stockholm: WWF.

Dauvergne, Peter (2001), *Loggers and Degradation in the Asia-Pacific: Corporations and Environmental Management*, Cambridge: Cambridge University Press.

Dauvergne, Peter (2005), 'The environmental challenge to loggers in the Asia-Pacific: corporate practices in informal regimes of governance', in David L. Levy and Peter J. Newell (eds), *The Business of Global Environmental Governance*, Cambridge, MA: MIT Press, pp. 169–96.

Dauvergne, Peter (2008), *The Shadows of Consumption: Consequences for the Global Environment*, Cambridge, MA: MIT Press.

deFontaubert, A.C. (1995), 'The politics of negotiation at the United Nations conference on straddling fish stocks and highly migratory fish stocks', *Ocean and Coastal Management*, **29** (1–3), 79–91.

DiMaggio, Paul J. (1991), 'Constructing an organizational field as a professional project: U.S. arts museums, 1920–1940', in W.W. Powell and P.J. DiMaggio (eds), *The New Institutionalism in Organizational Analysis*, Chicago, IL: University of Chicago Press, pp. 267–92.

DiMaggio, P.J. and W.W. Powell (1983), 'The iron cage revisited: institutional isomorphism and collective rationality in organizational fields', *American Sociological Review*, **48** (April), 147–60.

DiMaggio, Paul J. and Walter W. Powell (1991), 'Introduction', in W.W. Powell and P.J. DiMaggio (eds), *The New Institutionalism in Organizational Analysis*, Chicago, IL: University of Chicago Press, pp. 1–38.

Dingwerth, Klaus (2007), *The New Transnationalism. Transnational Governance and Democratic Legitimacy*, Basingstoke: Palgrave Macmillan.

Dingwerth, K. and P. Pattberg (2009), 'World politics and organizational fields: the case of transnational sustainability governance', *European Journal of International Relations*, **15** (4), 707–43.

Djelic, Marie-Laure and Kerstin Sahlin-Andersson (eds) (2006), *Transnational Governance: Institutional Dynamics of Regulation*, Cambridge: Cambridge University Press.

Drori, Gili, John Meyer and Hokyu Hwang (2006), *Globalization and Organization: World Society and Organizational Change*, Oxford: Oxford University Press.

Elliott, Christopher (1999), *Forest Certification: Analysis from a Policy*

Network Perspective, thesis no. 1965, Lausanne: École Polytechnique Fédérale de Lausanne.

Elliott, C. and R. Schlaepfer (2001), 'The advocacy coalition framework: application to the policy process for the development of forest certification in Sweden', *Journal of European Public Policy*, **8** (4), 642–61.

Elster, Jon (1989), *Nuts and Bolts for the Social Sciences*, Cambridge: Cambridge University Press.

ENDS (Environmental Data Services) (2003), 'Home office admits to "illegal" timber procurement', *ENDS Report*, 342 (July).

ENDS (2004a), 'PEFC timber scheme "inadequate" says DEFRA', *ENDS Report*, 358 (July).

ENDS (2004b), 'FSC rethink aims to boost labelled timber volumes', *ENDS Report*, 355 (August), 29.

ENDS (2005), 'DEFRA's approval of industry-certified timber blasted by green groups', *ENDS Report*, 368 (September).

Espach R. (2006), 'When is sustainable forestry sustainable? The Forest Stewardship Council in Argentina and Brazil', *Global Environmental Politics*, **6** (2), 55–84.

European Commission (2003), *Communication from the Commission to the Council and the European Parliament: Forest Law Enforcement, Governance and Trade (FLEGT), Proposal for an EU Action Plan*, EU document COM(2003)251 final, May 21, Brussels: European Commission.

FAO (Food and Agriculture Organization of the United Nations) (1995), *Code of Conduct for Responsible Fisheries*, Rome: FAO.

FAO (1998), *Report of the Technical Consultation on the Feasibility of Developing Non-discriminatory Technical Guidelines for Eco-labelling of Products from Marine Capture Fisheries*, 21-3 October, Rome: FAO.

FAO (1999), *Report of the 23rd Session of the Committee of Fisheries*, 15-19 February, FAO fisheries report no. 595, Rome: FAO.

FAO (2001), *Global Forest Resources Assessment 2000: Main Report*, Rome: FAO.

FAO (2002), *The State of World Fisheries and Aquaculture 2002*, Rome: FAO.

FAO (2005), *The FAO Guidelines for the Ecolabelling of Fish and Fisheries Products from Marine Capture Fisheries*, Rome: FAO.

FAO (2007), *The State of World Fisheries and Aquaculture 2006*, Rome: FAO.

FAO (2009), *The State of World Fisheries and Aquaculture 2008*, Rome: FAO.

Fearon, James and Alexander Wendt (2002), 'Rationalism v. constructivism: a skeptical view', in W. Carlsnaes, T. Risse and B. Simmons (eds), *Handbook of International Relations*, London: Sage, pp. 52–72.

Finnemore, M. (1996), 'Norms, culture, and world politics: insights from sociology's institutionalism', *International Organization*, **50** (2), 325–47.

Finnemore, M. and K. Sikkink (1998), 'International norms dynamics and political change', *International Organization*, **52** (4), 887–917.

Fisher, Carolyn, Francisco Aguilar, Puja Jawahar and Roger Sedjo (2005), *Forest Certification: Toward Common Standards?* discussion paper 05-10, Washington, DC: Resources for the Future.

Fowler, Penny and Simon Heap (2000), 'Bridging troubled waters: the Marine Stewardship Council', in Jem Bendell (ed.), *Terms for Endearment: Business, NGOs and Sustainable Development*, Sheffield: Greenleaf, pp. 135–48.

Franck, Thomas M. (1990), *The Power of Legitimacy Among Nations*, New York: Oxford University Press.

FSC (Forest Stewardship Council) (2000), *FSC Policy on Percentage Based Claims*, Oaxaca, Mexico: FSC International Center.

FSC (2002a), *Forest Stewardship Council A.C. By-laws*, document 1.1, Bonn: FSC International Center.

FSC (2002b), *FSC Principles and Criteria for Forest Stewardship, Approved 1993, Amended 1996, 1999, 2002*, FSC-STD-01-001 (V4-0) EN, Bonn: FSC International Center.

FSC (2003), *Response to the Rainforest Foundation Report 'Trading in Credibility'*, Bonn: FSC International Center.

FSC (2004), *Local Adaptation of Certification Body Generic Forest Stewardship Standards*, FSC-STD-20-003 (V2-1) EN, Bonn: FSC International Center.

FSC (2009), 'Milestones 2009', 23 December, *News & Notes*, **7** (11), Bonn: FSC International Center.

FSC Sweden (1998), *Swedish FSC Standard for Forest Certification*, Uppsala, Sweden: Swedish FSC Council.

Fung, Archon, Dara O'Rourke and Charles Sabel (eds) (2001), *Can We Put an End to Sweatshops? A New Democracy Forum on Raising Global Labor Standards*, Boston, MA: Beacon Press.

Garsten, Christina (2008), 'The United Nations – soft and hard: regulating social accountability for global business', in M. Boström and C. Garsten (eds), *Organizing Transnational Accountability*, Cheltenham, UK and Northampton, MA, USA: Edward Elgar, pp. 27–45.

Gelcich, S., G. Edward-Jones and M.J. Kaiser (2005), 'Importance of attitudinal differences among artisanal fisheres with respect to co-management and conservation of marine resources', *Conservation Biology*, **19**, 1–11.

George, Alexander L. and Andrew Bennett (2004), *Case Studies and*

Theory Development in the Social Sciences, Cambridge, MA: MIT Press.

Ghazoul, J. (2001), 'Barriers to biodiversity conservation in forest certification', *Conservation Biology*, **15**, 315–17.

Grant, R.W. and R.O. Keohane (2005), 'Accountability and abuses of Power in world politics', *American Political Science Review*, **99** (1), 1–15.

Gulbrandsen, L.H. (2003), 'The evolving forest regime and domestic actors: strategic or normative adaptation?', *Environmental Politics*, **12** (2), 95–114.

Gulbrandsen, L.H. (2004), 'Overlapping public and private governance: can forest certification fill the gaps in the global forest regime?', *Global Environmental Politics*, **4** (2), 75–99.

Gulbrandsen, L.H. (2005a), 'The effectiveness of non-state governance schemes: a comparative study of forest certification in Norway and Sweden', *International Environmental Agreements*, **5** (2), 125–49.

Gulbrandsen, L.H. (2005b), 'Mark of sustainability? Challenges for fishery and forestry eco-labeling', *Environment*, **47** (5), 8–23.

Gulbrandsen, L.H. (2006), 'Creating markets for eco-labelling: are consumers insignificant?', *International Journal of Consumer Studies*, **30** (5), 477–89.

Gulbrandsen, L.H. (2008), 'Accountability arrangements in non-state standards organizations: instrumental design and imitation', *Organization*, **15** (4), 563–83.

Gulbrandsen, L.H. (2009), 'The emergence and effectiveness of the Marine Stewardship Council', *Marine Policy*, **33** (4), 654–60.

Gulbrandsen, Lars H. and David Humphreys (2006), *International Initiatives to Address Tropical Timber Logging and Trade: A Report for the Norwegian Ministry of the Environment*, FNI Report 4/2006, Lysaker, Norway: Fridtjof Nansen Institute.

Gullison, R.E. (2003), 'Does forest certification conserve biodiversity?', *Oryx*, **37** (2), 153–65.

Gunningham, N. and J. Rees (1997), 'Industry self-regulation', *Law and Policy*, **17**, 363–414.

Haas, Peter M., Robert O. Keohane and Marc A. Levy (eds) (1993), *Institutions for the Earth: Sources of Effective International Environmental Protection*, Cambridge, MA: MIT Press.

Hall, Rodney Bruce and Thomas J. Biersteker (eds) (2002), *The Emergence of Private Authority in Global Governance*, Cambridge: Cambridge University Press.

Hardin, G. (1968), 'The tragedy of the commons', *Science*, **162**, 1243–8.

Hasenclever, Andreas, Peter Mayer and Volker Rittberger (1997), *Theories of International Regimes*, Cambridge: Cambridge University Press.

Haufler, Virginia (2001), *A Public Role for the Private Sector*, Washington, DC: Carnegie Endowment for International Peace.
Hellström, E. (2001), 'Conflict cultures – qualitative comparative analysis of environmental conflicts in forestry', *Silva Fennica Monographs*, 2, Helsinki: Finnish Society of Forest Science.
Highleyman, Scott, Amy Mathews Amos and Hank Cauley (2004), *An Independent Assessment of the Marine Stewardship Council*, draft report prepared for the Homeland Foundation, the Oak Foundation, and the Pew Charitable Trust, Bellingham, WA: Wildhavens.
Hoel, Alf Håkon (2006), 'An effective conservation tool? ecolabelling and fisheries', in Frank Aasche (ed.), *Primary Industries Facing Global Markets: The Supply Chains and Markets for Norwegian Food and Forest Products*, Oslo: Universitetsforlaget, pp. 347–73.
Hoel, Alf Håkon and I. Kvalvik (2006), 'The allocation of scarce natural resources: the case of fisheries', *Marine Policy*, **30**, 347–56.
Hoffman, A.J. (1999), 'Institutional evolution and change: environmentalism and the U.S. chemical industry', *Academy of Management Journal*, **42** (4), 351–71.
Honey, Martha (ed.) (2002), *Ecotourism and Certification: Setting Standards in Practice*, Washington, DC: Island Press.
Hooghe, Liesbet and Gary Marks (2001), *Multi-level Governance and European Integration*, Oxford: Rowman and Littlefield.
Howes, Rupert (2008), 'The Marine Stewardship Council programme', in Trevor Ward and Bruce Phillips (eds), *Seafood Ecolabelling: Principles and Practice*, Oxford: Wiley-Blackwell, pp. 81–105.
Humphreys, David (1996), *Forest Politics: The Evolution of International Cooperation*, London: Earthscan.
Humphreys, D. (1999), 'The evolving forest regime', *Global Environmental Change*, **9** (3), 251–4.
Humphreys, D. (2003), 'Life protective or carcinogenic challenge? Global forest governance under advanced capitalism', *Global Environmental Politics*, **3** (2), 40–55.
Humphreys, David (2006), *Logjam: Deforestation and the Crisis of Global Governance*, London: Earthscan.
Humphreys, J. (2002), 'MSC certification: funding support', *Samudra* (32), 23–5.
Håpnes, Arnodd and Stig Hvoslef (1999), *Levende Skog-standarden og den svenske FSC-standarden, en sammenlikning* [The Living Forest standard and the Swedish FSC standard, a comparison], World Wide Fund for Nature report 1999-5a, Oslo: WWF.
IFIR (International Forest Industry Roundtable) (2001), 'Proposing an international mutual recognition framework: report of the working

Group on Mutual Recognition between Credible Sustainable Forest Management Certification Systems and Standards', February, accessed at http://research.yale.edu/gisf/assets/pdf/tfd/IFIR_report.pdf.

ISEAL Alliance (2006), *ISEAL Code of Good Practice for Setting Social and Environmental Standards*, P005, public version 4, January, Oxford: ISEAL Alliance.

ITTO (International Tropical Timber Organization) (1992), *ITTO Guidelines for the Sustainable Management of Natural Tropical Forests*, Yokohama, Japan: ITTO.

Jacquet, J.L. and D. Pauly (2007), 'The rise of seafood awareness campaigns in an era of collapsing fisheries', *Marine Policy*, **31**, 308–13.

Jacquet, J.L. and D. Pauly (2008), 'Trade secrets: renaming and mislabeling of seafood', *Marine Policy*, **32**, 309–18.

Jordan, A., R.K.W. Wurzel and A.R. Zito (2003), '"New" environmental policy instruments: an evolution or a revolution in environmental policy?', *Environmental Politics*, **12** (1), 201–24.

Kaiser M.J. and G. Edward-Jones (2006), 'The role of ecolabelling in fisheries management and conservation', *Conservation Biology*, **20** (2), 392–8.

Kaplan, I. and B.J. McCay (2004), 'Cooperative research, co-management and the social dimensions of fisheries science and management', *Marine Policy*, **28**, 257–8.

Keck, Margaret E. and Kathryn Sikkink (1998), *Activist Beyond Borders: Advocacy Networks in International Politics*, Ithaca, NY: Cornell University Press.

Keohane, Robert O. (1984), *After Hegemony: Cooperation and Discord in the World Political Economy*, Princeton, NJ: Princeton University Press.

Keohane, Robert O. (1989), *International Institutions and State Power: Essays in International Relations Theory*, Boulder, CO: Westview Press.

Keohane, Robert O. (1993), 'The analysis of international regimes: towards a European-American research programme', in Volker Rittberger (ed.), *Regime Theory and International Relations*, Oxford: Clarendon Press, pp. 3–24.

Keohane, Robert O. (2003) 'Global governance and democratic accountability', in D. Held and M. Koenig-Archibugi (eds), *Taming Globalization: Frontiers of Governance*, Cambridge: Polity, pp. 130–59.

Keohane, Robert O. and Joseph S. Nye (1977), *Power and Interdependence*, Boston, MA: Little Brown.

King, Gary, Robert O. Keohane and Sidney Verba (1994), *Designing Social Inquiry: Scientific Inference in Qualitative Research*, Princeton, NJ: Princeton University Press.

Klein, D.B. (1997), 'Knowledge, reputation and trust by voluntary means', in D.B. Klein (ed.), *Reputation: Studies in the Voluntary Elicitation of Good Conduct*, Ann Arbor, MI: University of Michigan Press, pp. 1–14.

Klooster, D. (2005), 'Environmental certification of forests: the evolution of environmental governance in a commodity network', *Journal of Rural Studies*, **21**, 403–17.

Klooster, D. (2006), 'Environmental certification of forests in Mexico: the political ecology of a nongovernmental market intervention', *Annals of the Association of American Geographers*, **96**, 541–65.

Krasner, S. (1982), 'Structural causes and regime consequences: regimes as intervening variables', *International Organization*, **36** (2), 185–205.

Kratochwil, F. and J.G. Ruggie (1986), 'International organizations: a state of the art on an art of the state', *International Organization*, **40** (4), 753–75.

Krut, Riva and Harris Gleckman (1998), *ISO 14001: A Missed Opportunity for Sustainable Global Industrial Development*, London: Earthscan.

Kurien, J. (1996), 'Marine Stewardship Council: a view from the Third World', *Samudra* (15), 22–5.

Lack, M. and G. Sant (2001), 'Patagonian toothfish: are conservation and trade measures working?', *TRAFFIC Bulletin*, **19** (1).

Lambin, E.F. and H.J. Geist (2003), 'Regional differences in tropical deforestation', *Environment*, **45** (6), 22–36.

Leadbitter, D., G. Guillermo, and F. McGilvray (2006), 'Sustainable fisheries and the East Asian seas: can the private sector play a role', *Ocean and Coastal Management*, **49**, 662–75.

Leadbitter, D. and T. Ward (2007), 'An evaluation of systems for the integrated assessment of capture fisheries', *Marine Policy*, **31**, 458–69.

Lee, Daniel (2008), 'Aquaculture certification', in Trevor Ward and Bruce Phillips (eds), *Seafood Ecolabelling: Principles and Practice*, Oxford: Wiley-Blackwell, pp. 106–33.

Levy, M.A., O.R. Young and M. Zürn (1995), 'The study of international regimes', *European Journal of International Relations*, **1** (3), 267–330.

Lijphart, Arend (1971), 'Comparative politics and the comparative method', *The American Political Science Review*, **65** (3), 682–93.

Lindahl, Karin Beland (2001), *Behind the Logo: The Development, Standards and Procedures of the Forest Stewardship Council and the Pan European Forest Certification Scheme in Sweden*, Moreton-in-Marsh: FERN.

Lipschutz and Fogel (2002), '"Regulation for the rest of us?" Global civil society and the privatization of transnational regulations', in Rodney Bruce Hall and Thomas J. Biersteker (eds), *The Emergence of Private*

Authority in Global Governance, Cambridge: Cambridge University Press, pp. 115–40.

Living Forests (1998), *The Final Standard Documents from Living Forests, Report 9a-d*, Oslo: Living Forests.

Living Forests (2006), *Standard for Sustainable Forest Management in Norway*, Oslo: Living Forests.

Mäntyranta, H. (2002), *Forest Certification – An Ideal that Became an Absolute*, Vammalan Kirjapaino Oy: Metsälehti Kustannus.

March, James P. and Johan P. Olsen (1976), *Ambiguity and Choice in Organizations*, Bergen, Norway: Universitetsforlaget.

March, James P. and Johan P. Olsen (1989), *Rediscovering Institutions: The Organizational Basis of Politics*, New York: Free Press.

Mason, Michael (2005), *The New Accountability: Environmental Responsibility Across Borders*, London: Earthscan.

Mathew, S. (1998), 'Marine Stewardship Council: when sandals meet suits', *Samudra* (19), 31–5.

Mattli, W. and T. Büthe (2003), 'Setting international standards – technological rationality or primacy of power?', *World Politics*, **56** (1), 1–42.

May, Brendan, Duncan Leadbitter, Mike Sutton and Michael Weber (2003), 'The Marine Stewardship Council: background, rationale and challenges', in Bruce Phillips, Trevor Ward and Chet Chaffee (eds), *Eco-labelling in Fisheries: What is it All About?*, Oxford: Blackwell, pp. 14–33.

McAdam, Dough, Sidney Tarrow and Charles Tilly (2001), *Dynamics of Contention*, Cambridge: Cambridge University Press.

McDermott, C. and B. Cashore (2008), *Assessing USGBC's Forest Certification Policy Options: A Summary Report Prepared by the Yale Program on Forest Policy and Governance*, New Haven, CT: Yale Program on Forest Policy and Governance.

McDermott, C., E. Noah and B. Cashore (2008), 'Differences that matter? A framework for comparing environmental certification standards and government policies', *Journal of Environmental Policy and Planning*, **10**, 47–70.

McNichol, Jason (2002), *Contesting Governance in the Global Marketplace: A Sociological Assessment of NGO–Business Partnerships to Build Markets for Certified Wood Products*, Ph.D. dissertation, Berkeley, CA: University of California at Berkeley Department of Sociology.

Mearsheimer, J.J. (1995), 'The false promise of international institutions', *International Security*, **19** (3), 5–49.

Meidinger, E. (2006), 'The administrative law of global private-public regulation: the case of forestry', *The European Journal of International Law*, **17** (1), 47–87.

Meridian Institute (2001), *Comparative Analysis of the Forest Stewardship Council and Sustainable Forestry Initiative Certification Programs: Volume III, Description of the Sustainable Forestry Initiative Program*, Washington, DC: Meridian Institute.

Meyer, J.W., J. Boli, G.M. Thomas and F. Ramirez (1997), 'World society and the nation state', *American Journal of Sociology*, **103** (1), 144–81.

Meyer, J.W. and B. Rowan (1977), 'Institutionalized organizations: formal structure as myth and ceremony', *American Journal of Sociology*, **83** (2), 340–63.

Micheletti, Michele (2003), *Political Virtue and Shopping. Individuals, Consumerism, and Collective Action*, New York: Palgrave Macmillan.

Micheletti, Michele, Andreas Follesdal and Dietlind Stolle (eds) (2004), *Politics, Products and Markets*, New Brunswick, NJ: Transaction Press.

Miles, Edward L., Arild Underdal, Steinar Andresen, Jørgen Wettestad, Jon Birger Skjærseth and Elaine M. Carlin (2002), *Environmental Regime Effectiveness: Confronting Theory with Evidence*, Cambridge, MA: MIT Press.

Mol, Arthur P.J., Volkmar Lauber and Duncan Liefferink (2000), *The Voluntary Approach to Environmental Policy: Joint Environmental Policy-making in Europe*, Oxford: Oxford University Press.

MSC (Marine Stewardship Council) (2000a), 'World's first sustainable seafood products launched', press release, March 3, Marine Stewardship Council.

MSC (2000b), 'Support grows for the Marine Stewardship Council certification programme', press release, September 23, Marine Stewardship Council.

MSC (2001a), 'MSC announces new governance structure', press release, July 27, Marine Stewardship Council.

MSC (2001b), 'Alaska salmon processors/distributors pleased with eco-label', press release, November 13, Marine Stewardship Council.

MSC (2002), *MSC Principles and Criteria for Sustainable Fishing*, London: Marine Stewardship Council.

MSC (2005), 'MSC welcomes FAO guidelines on marine eco-labelling', press release, March 31, Marine Stewardship Council.

MSC (2006a), 'MSC not to develop aquaculture standard – MSC Board statement', press release, November 20, Marine Stewardship Council.

MSC (2006b), 'Leader in fishery certification and eco-labelling annouces 100 percent consistency with UN guidelines', press release, September 26, Marine Stewardship Council.

MSC (2006c), 'Wal-Mart sets 100 percent sustainable fish target for North America', press release, January 27, Marine Stewardship Council.

MSC (2006d), 'ASDA to switch to MSC certified fish in all its stores within 3–5 years', press release, March 28, Marine Stewardship Council.

MSC (2007), 'Objections panel for New Zealand hoki fishery concludes process', press release, September 17, Marine Stewardship Council.

MSC (2008a), 'Simpler, faster and more consistent: MSC launches new fisheries assessment methodology', press release, July 20, Marine Stewardship Council.

MSC (2008b), *Marine Stewardship Council Fisheries Asssessment Methodology and Guidance to Certification Bodies*, London: Marine Stewardship Council.

MSC (2008c), *Marine Stewardship Council Annual Report 2007/08*, London: Marine Stewardship Council.

Murphy, David F. and Jem Bendell (1997), *In the Company of Partners: Business, Environmental Groups and Sustainable Development Post-Rio*, Bristol: The Policy Press.

Nebel G., L. Quevedo, J.B. Jacobsen and F. Helles (2005), 'Development and economic significance of forest certification: the case of FSC in Bolivia', *Forest Policy and Economics*, **7**, 175–86.

Newell, P. (2005), 'Citizenship, accountability and community: the limits of the CSR agenda', *International Affairs*, **81** (3), 541–57.

Newell, Peter and Joanna Wheeler (eds) (2006), *Rights, Resources and the Politics of Accountability*, London: Zed Books.

Newsom D., V. Bahn and B. Cashore (2006), 'Does forest certification matter? An analysis of operation-level changes required during the SmartWood certification process in the United States', *Forest Policy and Economics*, **9**, 197–208.

Newsom Deanna and Daphne Hewitt (2005), *The Global Impact of SmartWood Certification*, New York: TREES Program, Rainforest Alliance.

Nussbaum, Ruth and Markku Simula (2005), *The Forest Certification Handbook*, 2nd edn, London: Earthscan.

OECD (Organisation for Economic Co-operation and Development) (2003), *Developing-country Access to Developed-country Markets under Selected Eco-labelling Programmes*, COM/ENV/TD(2003)30, Paris: OECD.

Olson, Mancur (1965), *The Logic of Collective Action*, Cambridge, MA: Harvard University Press.

Oosterveer, Peter (2005), *Global Food Governance*, Ph.D. dissertation, Wageningen, Netherlands: Wageningen University.

O'Riordan, B. (1997), 'Marine Stewardship Council: who's being seduced?', *Samudra* (18), 10–11.

O'Riordan, B. (1998), 'Sticky labels', *Samudra* (21), 22–5.

Orrego Vicuña, Francisco (2001), 'The international law of high seas fisheries: from freedom of fishing to sustainable use', in Olav Schram Stokke (ed.), *Governing High Seas Fisheries. The Interplay of Global and Regional Regimes,* Oxford: Oxford University Press, pp. 23–52.

Ostrom, Elinor (1990), *Governing the Commons: The Evolution of Institutions of Collective Action*, Cambridge: Cambridge University Press.

Overdevest, C. (2004), 'Codes of conduct and standard setting in the forestry sector: constructing markets for democracy?', *Relations Industrielles – Industrial Relations*, **59** (1), 172–97.

Overdevest, C. (2005), 'Treadmill politics, information politics, and public policy: toward a political economy of information', *Organization and Environment*, **18** (1), 72–90.

Ozinga, Saskia (2001), *Behind the Logo: An Environmental and Social Assessment of Forest Certification Schemes*, Moreton-in-Marsh: FERN.

Ozinga, Saskia (2004), *Footprints in the Forests: Current Practices and Future Challenges in Forest Certification*, Moreton-in-Marsh: FERN.

Pattberg, Philipp H. (2007), *Private Institutions and Global Governance: The New Politics of Environmental Sustainability*, Cheltenham, UK and Northampton, MA, USA: Edward Elgar.

Pearce, F. (2003), 'Can ocean friendly labels save dwindling stocks?', *New Scientist*, **178**, 5.

PEFC (Programme for the Endorsement of Forest Certification) (2004a), 'PEFC Council joins International Accreditation Forum', press release, 3 March, Luxembourg: PEFC Council.

PEFC (2004b), 'World Bank-WWF questionnaire hopelessly flawed – new approach offered by PEFC', press release, October 7, Luxembourg: PEFC Council.

PEFC (2006), *Rules for Standard Setting*, Geneva: PEFC Council.

PEFC (2007), *Certification and Accreditation Procedures*, Geneva: PEFC Council.

PEFC (2009), *Basis for Certification Schemes and Their Implementation*, Geneva: PEFC Council.

PEFC Sweden (Programme for the Endorsement of Forest Certification) (2000), *Swedish PEFC Technical Document I, 2000–2005*, Uppsala, Sweden: Swedish PEFC Council.

PEFC Sweden (2006), *Swedish PEFC Technical Document II, 2006–2011*, Uppsala, Sweden: Swedish PEFC Council.

Pellizzoni, Luigi (2004), 'Responsibility and environmental governance', *Environmental Politics*, **13** (3), 541–65.

Phillips, Bruce, Luis Bourillón and Mario Ramade (2008), 'Case study 2:

the Baja California, Mexico, lobster fishery', in Trevor Ward and Bruce Phillips (eds), *Seafood Ecolabelling: Principles and Practice*, Oxford: Wiley-Blackwell, pp. 259–68.

Phillips, Bruce, Trevor Ward, and Chet Chaffee (eds) (2003), *Eco-labelling in Fisheries: What is it All About?* Oxford: Blackwell.

Pierre, Jon (ed.) (2000), *Debating Governance: Authority, Steering, and Democracy*, Oxford: Oxford University Press.

Pierson, Paul (1993), 'When effect becomes cause: policy feedback and political change', *World Politics*, **45** (4), 595–628.

Pierson, Paul (2004), *Politics in Time: History, Institutions, and Social Analysis*, Princeton, NJ: Princeton University Press.

Ponte, S. (2008), 'Greener than thou: the political economy of fish ecolabeling and its local manifestations in South Africa', *World Development*, **36** (1), 159–75.

Ponte, S. and P. Gibbon (2005), 'Quality standards, conventions and the governance of global value chains', *Economy and Society*, **34** (1), 1–31.

Potts, T. and M. Haward (2007), 'International trade, eco-labelling, and sustainable fisheries – recent issues, concepts and practices', *Environment, Development and Sustainability*, **9** (1), 91–106.

Power, Michael (1997), *The Audit Society: Rituals of Verification*, Oxford: Oxford University Press.

Putz F. and C. Romero (2001), 'Biologists and timber certification', *Conservation Biology*, **15**, 313–14.

Ragin, Charles C. (1987), *The Comparative Method: Moving Beyond Qualitative and Quantitative Strategies*, Los Angeles, CA: University of California Press.

Rainforest Alliance (2007), 'Rainforest Alliance: highlights from the first 20 years 1986–2007', accessed at www.rainforest-alliance.org/about/documents/ra_timeline.pdf.

Rametsteiner, Ewald (1999), *Sustainable Forest Management Certification: Framework Conditions, Systems Designs and Impact Assessment*, Vienna: MCPFE Liasion Unit.

Rametsteiner, J. and M. Simula (2003), 'Forest certification – an instrument to promote sustainable forest management?', *Journal of Environmental Management*, **67** (1), 87–98.

Raynolds, Laura and Douglas Murray (2007), 'Fair trade: contemporary challenges and future prospects', in Laura Raynolds, Douglas Murray and John Wilkinson (eds), *Fair Trade: The Challenges of Transforming Globalization*, London: Routledge, pp. 223–34.

Raynolds, Laura, Douglas Murray and John Wilkinson (eds) (2007), *Fair Trade: The Challenges of Transforming Globalization*, London: Routledge.

Reinecke, Wolfgang (1998), *Global Public Policy*, Washington, DC: Brookings Institution.

Rhodes, R.A.W. (1996), 'The new governance: governing without government', *Political Studies*, **44**, 652–67.

Rhodes, R.A.W. (1997), *Understanding Governance: Policy Networks, Governance, Reflexivity and Accountability*, Maidenhead: Open University Press.

Risse, Thomas, Stephen Ropp and Kathryn Sikkink (eds) (1999), *The Power of Human Rights: International Norms and Domestic Change*, Cambridge: Cambridge University Press.

Risse-Kappen, Thomas (ed.) (1995), *Bringing Transnational Relations Back In: Non-state Actors, Domestic Structures and International Institutions*, Cambridge: Cambridge University Press.

Romzek, B.S. and M.J. Dubnick (1987), 'Accountability in the public sector: lessons from the Challenger tragedy', *Public Administration Review*, **47**, 227–38.

Rosendal, G.K. (2001), 'Overlapping international regimes: the case of the Intergovernmental Forum on Forests (IFF) between climate change and biodiversity', International Environmental Agreements, **1** (4), 447–68.

Rosenau, J.N. (1995), 'Governance in the twenty-first century', *Global Governance*, **1** (1), 13–43.

Rosenau, James N. (1997), *Along the Domestic-foreign Frontier: Exploring Governance in a Turbulent World*, Cambridge: Cambridge University Press.

Rosenau, James N. (2000), 'Change, complexity, and governance in a globalizing space', in Jon Pierre (ed.), *Debating Governance: Authority, Steering, and Democracy*, Oxford: Oxford University Press, pp. 167–200.

Rosenau, James N. (2003), *Distant Proximities: Dynamics beyond Globalization*, Princeton, NJ: Princeton University Press.

Rosenau, James N. and Ernst-Otto Czempiel (eds) (1992), *Governance Without Government: Order and Change in World Politics*, Cambridge: Cambridge University Press.

Rosoman, Grant, Judy Rodrigues and Anna Jenkins (2008), *Holding the Line with the FSC*, Amsterdam: Greenpeace International.

Ruggie, John G. (1983), 'International regimes, transaction costs, and change: embedded liberalism in the postwar economic order', in Stephen D. Krasner (ed.), *International Regimes*, Ithaca, NY: Cornell University Press, pp. 195–231.

Sabatier, Paul (1998), 'The Advocacy Coalition Framework: revisions and relevance for Europe', *Journal of European Public Policy*, **5** (1), 98–130.

Sabatier, Paul and Hank Jenkins-Smith (1999), 'The Advocacy Coalition Framework: an assessment', in Paul Sabatier (ed.), *Theories of the Policy Process*, Boulder, CO: Westview Press, pp. 117–66.

Sahlin-Andersson, Kerstin (1996), 'Imitating by editing success: the construction of organizational fields', in B. Czarniawska and G. Sevón (eds), *Translating Organizational Change*, Berlin: de Gruyter, pp. 69–92.

Sahlin-Andersson, Kerstin and Lars Engwall (eds) (2002), *The Expansion of Management Knowledge: Carriers, Flows and Sources*, Stanford, CA: Stanford University Press.

Samudra (1996), 'Going green about the gills', editorial, *Samudra*, (15), 1.

Sasser, E., A. Prakash, B. Cashore and G. Auld (2006), 'Direct targeting as an NGO political strategy: examining private authority regimes in the forestry sector', *Business and Politics*, **8**, 1–32.

Scott, W. Richard (2001), *Institutions and Organizations*, 2nd edn, Thousand Oaks, CA: Sage.

Scott, W. Richard (2003), *Organizations: Rational, Natural and Open Systems*, 5th edn, Upper Saddle River, NJ: Prentice Hall.

Seafood Watch Evaluation (2004), *Summary Report*, Saltspring Island, BC: Quadra Planning Consultants Ltd. and Galiano Institute for Environmental and Social Research.

Sedjo R.A. and D. Botkin (1997), 'Using forest plantations to spare natural forests', *Environment*, **39**, 14–22.

SEPA (Swedish Environmental Protection Agency) (2004), *Sweden's Environmental Objectives: Are we Getting There? De facto 2004*, Stockholm: SEPA.

Snidal, D. (1985), '"Coordination versus prisoners" dilemma: implications for international relations and regimes', *American Political Science Review*, **79** (4), 923–42.

Spar, D.L. (1998), 'The spotlight and the bottom line: how multinationals export human rights', *Foreign Affairs*, **77**, 7–12.

Stern, Nicholas (2006), *Stern Review: The Economics of Climate Change*, London: Her Majesty's Treasury.

Stokke, Olav Schram (1997), 'Regimes as governance systems', in Oran R. Young (ed.), *Global Governance: Drawing Insights from the Environmental Experience*, Cambridge, MA: MIT Press, pp. 27–64.

Stokke, Olav Schram (2001), 'Introduction', in Olav Schram Stokke (ed.), *Governing High Seas Fisheries: The Interplay of Global and Regional Regimes*, Oxford: Oxford University Press, pp. 1–19.

Stokke, O.S. (2004), 'Labelling, legalisation and sustainable management of forestry and fisheries', paper presented at the fifth Pan-European International Relations Conference, September 9-11, The Hague, Netherlands.

Strange, S. (1982), 'Cave! Hic Dragones: a critique of regime analysis', *International Organization*, **36** (2), 479–510.

Strange, Susan (1996), *The Retreat of the State: The Diffusion of Power in the World Economy*, Cambridge: Cambridge University Press.

Sutton, David (2003), 'An unsatisfactory encounter with the MSC – a conservation perspective', in Bruce Phillips, Trevor Ward and Chet Chaffee (eds), *Eco-labelling in Fisheries: What is it All About?* Oxford: Blackwell, pp. 114–19.

Sutton, M. (1996), 'Marine Stewardship Council: new hope for marine fisheries', *Samudra*, (15), 15–18.

Sutton, M. (1998), 'Marine Stewardship Council: an appeal for co-operation', *Samudra*, (19), 26–30.

Sutton, M. and C. Whitfield (1996), 'Marine Stewardship Council: a powerful arrow in the quiver', *Samudra* (16), 33–4.

Sverdrup-Thygeson, A., P. Borg and E. Bergsaker (2008), 'A comparison of biodiversity values in boreal forest regeneration areas before and after certification', *Scandinavian Journal of Forest Research*, **23** (3), 236–43.

Swedish Forest Agency (2008), *Statistical Yearbook of Forestry 2008*, Jönköping, Sweden: Swedish Forest Agency.

Synnott, Timothy (2005), 'Some notes on the early years of FSC', accessed at www.fsc.org/fileadmin/web-data/public/document_center/publications/Notes_on_the_early_years_of_FSC_by_Tim_Synnott.pdf.

Sæther, B. (1998), 'Environmental improvements in the Norwegian pulp and paper industry – from place and government to space and market', *Norsk geografisk Tidsskrift*, **52**, 181–94.

Teisl, M.F., B. Roe and R.L. Hicks (2002), 'Can eco-labels tune a market? Evidence from dolphin safe labelling', *Journal of Environmental Economics and Management*, **43**, 339–59.

Tollefson, Chris, Fred Gale and David Haley (2008), *Setting the Standard: Certification, Governance, and the Forest Stewardship Council*, Vancouver, BC: UBC Press.

Thrane, M., F. Ziegler and U. Sonesson (2009), 'Eco-labelling of wild-caught seafood products', *Journal of Cleaner Production*, **17**, 416–23.

UN (United Nations) (1992a), *United Nations Convention on Biological Diversity*, Nairobi: United Nations Environment Programme.

UN (1992b), *Non-legally Binding Authoritative Statement of Principles for a Global Consensus on the Management, Conservation and Sustainable Development of All Types of Forests*, A/CONF.151/26 (Vol. III).

UN (1997), *Kyoto Protocol to the United Nations Framework Convention on Climate Change*, Kyoto, 11 December.

UNCTAD (United Nations Conference on Trade and Development)

(2007), *Workshop on Environmental Requirements and Market Access for Developing Countries: How to Turn Challenges into Opportunities*, Geneva: UNCTAD.

Underdal, A. (1992), 'The concept of regime "effectiveness"', *Cooperation and Conflict*, **27** (3), 227–40.

Underdal, Arild (2002), 'One question, two answers', in Edward L. Miles, Arild Underdal, Steinar Andresen, Jørgen Wettestad, Jon Birger Skjærseth and Elaine M. Carlin, *Environmental Regime Effectiveness: Confronting Theory with Evidence*, Cambridge, MA: MIT Press, pp. 3–45.

UNECE/FAO (United Nations Economic Commission for Europe/FAO) (1999), *The Status of Forest Certification in the ECE Region, Geneva Timber and Forest Certification Papers*, ECE/TIM/DP/14, New York and Geneva: UN.

UNECE/FAO (2007), *Forest Products Annual Market Review 2006–2007*, ECE/TIM/SP/22, New York and Geneva: UN.

UNECE/FAO (2009), *Forest Products Annual Market Review 2008–2009*, ECE/TIM/SP/24, New York and Geneva: UN.

UNEP (United Nations Environment Program) (1995), *Global Biodiversity Assessment*, Cambridge: Cambridge University Press.

UNEP (2002), *Forest Biological Diversity: Convention on Biological Diversity: 6th Conference of the Parties*, CBD/COP/6/Decision VI/22.

UNFCCC (United Nations Framework Convention on Climate Change) (2007), *Report of the Conference of the Parties on Its Thirteenth Session, Held in Bali from 3 to 15 December 2007*, FCCC/CP/2007/6/Add.1.

Unilever (2005), 'Environmental and social report 2005', accessed at www.unilever.com/ourvalues/environmentandsociety/env_social_report.

Vallejo, Nancy and Pierre Hauselmann (2001), 'PEFC – an analysis', World Wide Fund for Nature discussion paper, January.

Vogel, David (1995), *Trading Up: Consumer and Environmental Regulation in a Global Economy*, Cambridge, MA: Harvard University Press.

Vogel, David (2005), *The Market for Virtue: The Potential and Limits of Corporate Social Responsibility*, Washington, DC: Brookings Institution Press.

Vogel, D. (2008), 'Private global business regulation', *Annual Review of Political Science*, **11**, 261–82.

Waltz, Kenneth N. (1979), *Theory of International Politics*, New York: Random House.

Wapner, Paul (1996), *Environmental Activism and World Civic Politics*, Albany, NY: SUNY Press.

Ward, T.J. (2008), 'Barriers to biodiversity conservation in marine fishery certification', *Fish and Fisheries*, **9**, 167–77.

Ward, Trevor J. and Bruce Phillips (2008), 'Anecdotes and lessons of a decade', in Trevor Ward and Bruce Phillips (eds), *Seafood Ecolabelling: Principles and Practice*, Oxford: Wiley-Blackwell, pp. 416–35.
Watson, R. and D. Pauly (2001), 'Systematic distortions in world fisheries catch trends', *Nature*, **414**, 536–8.
WBCSD (World Business Council for Sustainable Development) (2003), 'Forest certification systems and the "Legitimacy" Thresholds Model (LTM)', WBCSD discussion paper, Geneva, Switzerland.
Weir, Anne (2000), 'Meeting social and environmental objectives through partnership: the experience of Unilever', in Jem Bendell (ed.), *Terms for Endearment: Business, NGOs and Sustainable Development*, Sheffield: Greenleaf, pp. 118–24.
World Bank (2002), *The World Bank Operational Manual: Operational Policies*, OP 4.36, Washington, DC: World Bank Group.
WWF (World Wide Fund for Nature) (2009), 'The global forest and trade network', accessed at http://gftn.panda.org/.
Yin, Robert K. (2003), *Case Study Research: Design and Methods*, 3rd edn, London: Sage.
Young, O.R. (1986), 'International regimes: toward a new theory of institutions', *World Politics*, **39**, 104–22.
Young, Oran R. (1989), *International Cooperation: Building Regimes for Natural Resources and the Environment*, Ithaca, NY: Cornell University Press.
Young, Oran R. (1999), *Governance in World Affairs*, Ithaca, NY: Cornell University Press.
Young, Oran R. and Marc A. Levy (1999), 'The effectiveness of international environmental regimes', in Oran R. Young (ed.), *The Effectiveness of International Environmental Regimes: Causal Connections and Behavioral Mechanisms*, Cambridge, MA: MIT Press, pp. 1–32.
Young, Oran R. and Gail Osherenko (eds) (1993), *Polar Politics: Creating International Environmental Regimes*, Ithaca, NY: Cornell University Press.

Index